Prevention of Injuries and Overuse in Sports

Hermann O. Mayr • Stefano Zaffagnini
Editors

Prevention of Injuries and Overuse in Sports

Directory for Physicians, Physiotherapists, Sport Scientists and Coaches

 Springer

Editors
Hermann O. Mayr
Schoen Clinic Munich Harlaching
Munich
Germany

ESSKA ASBL
Centre Médical
Fondation Norbert Metz
76, rue d'Eich
1460 Luxembourg
Luxembourg

Stefano Zaffagnini
Istituti Ortopedico Rizzoli
Clinica Ortopedica Traumatologica I
Bologna
Italy

ISBN 978-3-662-51604-1   ISBN 978-3-662-47706-9 (eBook)
DOI 10.1007/978-3-662-47706-9

Springer Heidelberg New York Dordrecht London
© ESSKA 2016
Softcover re-print of the Hardcover 1st edition 2016

Printed on acid-free paper

Springer-Verlag GmbH Berlin Heidelberg is part of Springer Science+Business Media (www.springer.com)

Preface

The present focus of the medical profession and of physiotherapy in sports medicine is to treat sports injuries and overload syndromes due to sports. In sports medicine, usually surgical and rehabilitation issues are handled.

According to the European Association for Injury Prevention and Safety Promotion (EuroSafe), about 6.1 million people in Europe are treated in hospitals for a sports injury annually. About 420,000 of these cases have to be admitted for further treatment. The estimated number of hospital-treated sports injuries is 2.5 million in team ball sports, followed by 500,000 cases in ice or snow sports and the same number in acrobatic sports. Sixty per cent of sports injuries result from participation in organised sports. The number of sports injuries is increasing, probably due to increased participation in sports (EuroSafe, Injuries in the European Union, Report on injury statistics 2008–2010, Amsterdam, 2013, ISBN: 978-90-6788-464-8).

Medical treatment due to overload in sports currently lacks statistical analysis. Overload frequently results in a need for medical treatment or leads to injury.

Very popular sports such as football and alpine skiing are particularly relevant because of the injury frequency. The treatment costs are socioeconomically important.

Individual suffering and remaining disabilities after injury are a tragic aspect. Athlete downtime in competitive sports weakens the teams and affects athletes' careers.

It is therefore important to consider prevention more carefully in the future. The Sports Committee of ESSKA has set itself the task to discuss and to describe the main aspects of prevention. Prevention of injury and overuse in sports is not exclusively in the competence of physicians and physiotherapists. Therefore, the Sports Committee has partnered with a large group of experienced sports scientists, physiotherapists with experience in high-performance sports and coaches from high-performance sports to deliver this book on the prevention of injury and overuse in sports. This book cannot definitely answer all the questions on prevention. However, it should pave the way in cooperative prevention research as a directory for physicians, sports scientists and physiotherapist coaches.

The authors hope to provide help in the prevention of injuries and overuse in sports with this book.

Munich, Germany   Hermann O. Mayr
January, 2016

Contents

Contributors

Peter Angele Department of Trauma Surgery, FIFA Medical Centre of Excellence, University Medical Centre Regensburg, Regensburg, Germany

Sporthopaedicum Straubing/Regensburg, Regensburg, Germany

Martin Auracher Rehabilitation Center, Osteo Zentrum Schliersee, Schliersee, Germany

Mario Bizzini, PhD, PT F-MARC (FIFA Medical Assesment and Research Centre), Schulthess Clinic, Zurich, Switzerland

Yoann Bohu, MD Department of Orthopaedic Surgery, Clinique du Sport Paris, Paris, France

Department of Orthopaedic Surgery, Institut de l'Appareil Locomoteur Nollet, Paris, France

Ann M.J. Cools, PhD, PT Department of Rehabilitation Sciences and Physiotherapy, Ghent University, Ghent, Belgium

Tommaso Roberti di Sarsina, MD Clinica Ortopedica e Traumatologica II, Laboratorio di Biomeccanica ed Innovazione Tecnologica, Istituto Ortopedico Rizzoli, Bologna, Italy

Pascal Edouard Department of Clinical and Exercise Physiology, Sports Medicine Unity, University of Saint-Etienne, Faculty of medicine, Saint-Etienne, France

Laboratory of Exercise Physiology (LPE EA 4338),University of Lyon, Saint-Etienne, France

Joao Espregueira-Mendes Clínica do Dragão – Espregueira Mendes Sports Centre – FIFA Medical Centre of Excellence, Porto, Portugal

Christian Fink, MD Research Unit for Orthopedic Sports Medicine and Injury Prevention, Institute of Sports Medicine, Alpine Medicine and Health Tourism, UMIT – The Health and Life Sciences University, Innsbruck, Austria

FIFA Medical Centre of Excellence Innsbruck/ Tirol, Innsbruck, Austria

Gelenkpunkt – Center for Sports and Joint Surgery, Innsbruck, Austria

Felix Fischer, MSc Research Unit for Orthopedic Sports Medicine and Injury Prevention, Institute of Sports Medicine, Alpine Medicine and Health Tourism, UMIT – The Health and Life Sciences University, Hall in Tirol, Austria

FIFA Medical Centre of Excellence, Innsbruck/Tirol, UMIT – The Health and Life Sciences University, Hall in Tirol, Austria

Peter Gföller, MD FIFA Medical Centre of Excellence Innsbruck/Tirol, Innsbruck, Austria

Gelenkpunkt – Center for Sports and Joint Surgery, Innsbruck, Austria

Alberto Grassi, MD Clinica Ortopedica e Traumatologica II, Laboratorio di Biomeccanica ed Innovazione Tecnologica, Istituto Ortopedico Rizzoli, Bologna, Italy

Caroline Hepperger Research Unit for Orthopedic Sports Medicine and Injury Prevention, Institute of Sports Medicine, Alpine Medicine and Health Tourism, UMIT – The Health and Life Sciences University, Hall in Tirol, Austria

FIFA Medical Centre of Excellence Innsbruck/Tirol, Innsbruck, Austria

Gelenkpunkt – Center for Sports and Joint Surgery, Innsbruck, Austria

Mirco Herbort, MD Department of Traumatology, Hand- and Reconstructive Surgery, University Hospital Muenster (UKM), Münster, Germany

Serge Herman, MD Department of Orthopaedic Surgery, Clinique du Sport Paris, Paris, France

Department of Orthopaedic Surgery, Institut de l'Appareil Locomoteur Nollet, Paris, France

Helmut Hoffmann, MD, MOS, MSM Eden Reha, FIFA Medical Centre of Excellence, Donaustauf, Germany

Henrique Jones, MD, MOS, MSM Montijo Orthopedic and Sports Medicine Clinic, Lusofona University, Montijo, Lisbon, Portugal

Mary Jones FIFA Medical Centre of Excellence, Fortius Clinic, London, London, UK

Val Jones, PT Sheffield Shoulder and Elbow Unit, Department of Orthopaedic, Northern General Hospital, Sheffield, UK

Shahnaz Klouche Department of Orthopaedic Surgery, Clinique du Sport Paris, Paris, France

Department of Orthopaedic Surgery, Institut de l'Appareil Locomoteur Nollet, Paris, France

Werner Krutsch Department of Trauma Surgery, FIFA Medical Centre of Excellence, University Medical Centre Regensburg, Regensburg, Germany

Joris R. Lansdaal, MD Department of Orthopaedic Surgery, Central Military Hospital/University Medical Centre, Utrecht, The Netherlands

Nicolas Lefevre, MD Department of Orthopaedic Surgery, Clinique du Sport Paris, Paris, France

Department of Orthopaedic Surgery, Institut de l'Appareil Locomoteur Nollet, Paris, France

Antonio Maestro, MD, PhD FREMAP, Gijón, Spain

Antoine Marsaudon Department of Economics, Paris School of Economics (Université Paris 1) and Hospinnomics, Centre d'Economie de la Sorbonne, Paris, France

Hermann O. Mayr Department of Orthopaedics and Traumatology, University Hospital Freiburg, Freiburg, Germany

Schoen Clinic Munich Harlaching, Munich, Germany

Jacques Menetrey, MD Swiss Olympic Medical Center, Hôpitaux Universitaires de Genève, Faculté de médecine de Genève, Genève, Switzerland

Centre de Médecine de l'appareil locomoteur et du Sport – HUG, Hôpitaux Universitaires de Genève, Faculté de médecine de Genève, Genève, Switzerland

Département de chirurgie, Hôpitaux Universitaires de Genève, Faculté de médecine de Genève, Genève, Switzerland

Max Merkel Osteo Zentrum Schliersee, Schliersee, Germany

Florian Müller Physiotherapeutic Practice, Alternative Therapy, Bayrischzell, Germany

Federico Raggi, MD Clinica Ortopedica e Traumatologica II, Laboratorio di Biomeccanica ed Innovazione Tecnologica, Istituto Ortopedico Rizzoli, Bologna, Italy

Margherita Ricci, MD Isokinetic Medical Group, FIFA Medical Centre of Excellence, Bologna, Italy

Lise Rochaix Department of Economics, Paris School of Economics (Université Paris 1) and Hospinnomics, Centre d'Economie de la Sorbonne, Paris, France

Manuel Rodríguez, MD Sport and Health Center, Balneary Las Caldas, Oviedo, Spain

Christoph Schabbehard Dipl. Sportlehrer, Lautrach, Germany

Elvire Servien, MD, PhD Department of Orthopaedic Surgery, Hopital de la Croix-Rousse, Centre Albert Trillat, Lyon University, Lyon, France

Jorge Silvério Clínica do Dragão – Espregueira Mendes Sports Centre – FIFA Medical Centre of Excellence, Porto, Portugal

Patricia Thoreux Service de Chirurgie Orthopédique, APHP – Hôpital Avicenne, Université Paris 13, Bobigny, France

Institut de Biomécanique Humaine Georges Charpak, Arts et Métiers Paris Tech. 151 boulevard de l'Hôpital, Paris, France

Département médicalINSEP (Institut National du Sport, de l'Expertise et de la Performance), Paris, France

Bernd Thurner Dipl. Sportlehrer, Prevention Programs, Therapy- and Training Center Friedberg, Thomas-, Friedberg, Germany

Xavier Torrallardona, MSc Sport and Health Center, Balneary Las Caldas, Oviedo, Spain

Michel P.J. van den Bekerom, MD Department of Orthopaedic Surgery, Onze Lieve Vrouwe Gasthuis, Amsterdam, The Netherlands

Gorka Vázquez, Pt, DO Human Anatomy and Embryology Department, University of Oviedo, Oviedo, Spain

Francesco Della Villa, MD Isokinetic Medical Group, FIFA Medical Centre of Excellence, Bologna, Italy

Stefano Della Villa, MD Isokinetic Medical Group, FIFA Medical Centre of Excellence, Bologna, Italy

Karlheinz Waibel German Ski Federation, Science and Technology, Planegg, Germany

Andrew Williams FIFA Medical Centre of Excellence, Fortius Clinic, London, UK

Stefano Zaffagnini Clinica Ortopedica e Traumatologica II, Laboratorio di Biomeccanica ed Innovazione Tecnologica, Istituto Ortopedico, Bologna, Italy

General Considerations on Sports-Related Injuries

1

Patricia Thoreux, Pascal Edouard,
Antoine Marsaudon, and Lise Rochaix

P. Thoreux (✉)
Service de Chirurgie Orthopédique,
APHP – Hôpital Avicenne, Sorbonne Paris Cité,
Université Paris 13, 125 route de Stalingrad,
Bobigny 93009, France

Institut de Biomécanique Humaine Georges Charpak,
Arts et Métiers Paris Tech. 151 boulevard de
l'Hôpital, Paris 75013, France

Département médical,
INSEP (Institut National du Sport, de l'Expertise et
de la Performance), 11 avenue du Tremblay,
75012 Paris, France
e-mail: patricia.thoreux@avc.aph.fr

P. Edouard
Department of Clinical and Exercise Physiology,
Sports Medicine Unity, University of Saint-Etienne,
Faculty of medicine, Saint-Etienne
F-42055, France

Laboratory of Exercise Physiology (LPE EA 4338),
University of Lyon, Saint-Etienne
F-42023, France
e-mail: pascal.edouard42@gmail.com

A. Marsaudon • L. Rochaix
Department of Economics, Paris School of
Economics (Université Paris 1) & Hospinnomics,
Centre d'Economie de la Sorbonne,
106-112, Boulevard de l'Hôpital,
Paris F-75647, France
e-mail: antoine.marsaudon@gmail.com;
lise.rochaix@psemail.eu

Key Points
1. The epidemiological data collection represents the first step of the injury prevention sequence.
2. It is mandatory to determine the risk factors and injury mechanisms and to separate the intrinsic factors on which there is little to do and the extrinsic factors (like equipment, environment and ways of training).
3. The incidence of injuries is greatly different depending on the kind of sports, team sports are more traumatic than individual sports.
4. Socio economics of sports related injuries, both direct or indirect costs, are important to take into account.
5. Enhancing health through practising sports is very important to promote and has to be facilitated by a good knowledge of injuries prevention methods.

1.1 Introduction

The European Union has been actively supporting the development of regular physical activity for several health reasons [65] (COM

© ESSKA 2016
H.O. Mayr, S. Zaffagnini (eds.), *Prevention of Injuries and Overuse in Sports: Directory for Physicians, Physiotherapists, Sport Scientists and Coaches*, DOI 10.1007/978-3-662-47706-9_1

2007)[1,2]. The Eurobarometer,[3] published in 2013, provides statistics of sports participation and indicates that on average 58 % of Europeans engage in physical activity or sports. Nevertheless, the number of persons reporting never practicing sports or physical activity increased by 3 % since 2009 (39–42 %). In 2010, in France, two thirds of the population aged 15 and over report having practiced, over the last 12 months, one or more physical activities or sports at least once a week.[4]

Given the very large number of people practicing physical activity or sports, a broad economic and health perspective is needed on this issue. The lack (or excess) of physical activity or sports can generate significant direct and indirect economic costs. This is mainly due to the increase in health-care costs, longer sick leaves, or the increase in premature deaths. The costs borne by European Member States in relation to the practice of sports and physical activity mainly come from the increase in hospitalization rates (direct costs) and from productivity losses (indirect costs). But the practice of physical activity, by reducing morbidity, can also have large benefits, in particular for diseases in relation to obesity, such as type II diabetes or any kind of chronic diseases. Currently, the risk/benefit balance seems to be largely in favor of physical activity or sports practice [2, 7, 29, 67].

Whatever you consider competitive practice or physical activity as part of treatment for chronic diseases, injury represents a major problem in sports and physical activities given the consequences for the athlete himself at short, middle, or long term, for his entourage (coaches, leaders, family, physicians, etc.), and for society (financial, economical, absence at work, media, etc.). From an athlete's point of view, sports injury can of course lead to bad sports perfor-

mance and also to nonparticipation in a sports season and maybe to end of career; moreover, injury can lead to long-term consequences and/or disability. From the clinical practice point of view, sports injury represents an important part of sports medicine physicians and/or a health professional's work. Sports medicine staffs play (and have to play) an important role in the management of sports injury from the acute management and the diagnosis to the rehabilitation and the return to sports of an athlete. In case of injury, an athlete asks to be treated as soon as possible to quickly return to sports, with greater performance! The management of sports injury (as primary prevention, treatment, and secondary prevention) is one of the main roles of sports physicians to protect the health of athletes[5] [37, 41, 42, 48, 57]. From the scientific research point of view, injury is also of interest since 31824 articles were found on the PubMed database with *sports injury* keywords (at the 24/03/15). Thus, studying sports injuries is of interest for all stakeholders around the athletes.

Sports medicine should be based on evidence and be an evidence-based sports medicine. Pressures by stakeholders around athletes, such as coaches, media, and financial and political stakeholders, make the need of perfect health of athletes. Moreover, the current medico-economical problem requires that we have a thorough knowledge of our practices. In this context, sports injury management and even more sports injury prevention should be based on the best available evidence.

In addition, some major injuries lead to socioeconomic impacts, for example, knee injuries (especially during ski season or football practice) or ankle sprains. Sports-related or physical activity-related injuries lead to non-negligible direct and indirect costs which are difficult to evaluate. This latter consequence is important, since it reinforces the importance of prevention and includes as active partners the financial stakeholders.

In order to go ahead in sports injury prevention, it is of interest to better understand sports injury. Sports injury can be described by its location, type,

[1] Commission des Communautés européennes. Livre blanc sur le sport. COM(2007) 391 final, Bruxelles, 11 juillet 2007. http://ec.europa.eu/sport/white-paper/doc/wp_on_sport_fr.pdf

[2] World Health Organization (WHO) (2002). Reducing Risks, Promoting Healthy Life: World Health Report

[3] Commission Européenne: http://europa.eu/rapid/press-release_MEMO-14-207_fr.htm

[4] http://www.sports.gouv.fr/IMG/archives/pdf/Stat-Info_01-11_decembre2010.pdf

[5] International Olympic Committee. Olympic Movement Medical Code. In force as from 1 October 2009 http://www.olympic.org/PageFiles/61597/Olympic_Movement_Medical_Code_eng.pdf (accessed 15 Dec 2014)

mode of onset, causes, mechanisms, and severity [11, 56]. There are basically two modes of onset of injuries: sudden injuries and gradual injuries. A sudden onset incident refers to an episode where the experienced distress or disability developed within minutes, seconds, or less, while a gradual onset incident refers to an episode that developed within hours, days, or more [64]. In addition, there are also basically two causes of injuries: traumatic (macro-trauma) and overuse (microtrauma). Traumatic injuries are mainly acute with a sudden mode of onset and caused by an identifiable single external transfer of energy (a single traumatic event). Common examples include shoulder dislocations, wrist fractures, ankle sprains, Achilles rupture, and hamstring muscle strains. Overuse injuries refer to a condition to which no identifiable single external transfer of energy can be associated; multiple accumulative bouts of energy transfer could result in this kind of injury [64]. Overuse injuries can occur with sudden-onset injuries (e.g., tendon tears). Overuse injuries with gradual onset are more subtle and usually occur over time. They are the result of repetitive microtrauma to every anatomical structure of the neuromusculoskeletal system such as the bone, tendon, joint, or muscle. Common examples include tennis elbow (lateral epicondylitis), swimmer's shoulder (rotator cuff tendinitis and impingement with or without hyperlaxity), runner's knee, jumper's knee (infrapatellar tendinitis), Achilles tendinitis, shin splints, stress fractures in runners or dancers, etc. Overuse injuries occur when tissue adaptation fails: the human body and all the anatomical structures such as the bone, tendon, joint, and muscle have a physiological capacity to adapt its properties with a remodeling process depending on constraints or stresses (gravity, activity, loads, etc.). This remodeling process is an association of breaking down and building up (e.g., as the osseous callus). But sometimes, in certain circumstances, the process is exceeded; this can happen with the change in physical activity, when you have just started practicing a physical activity and/or if you increase your training workload in volume and/or intensity. Some favorable circumstances can usually be found and can be divided into intrinsic (as anatomical malalignment like genu valgum or varum, abnormalities of plantar sole) and extrinsic factors (as training errors, modification of surfaces, or footwear).

In other respects, we have to define the term of overload, which has to be clearly distinguished from the term of overuse [54]. The principle of overload states that a greater than normal stress or load on the body is required for training adaptation to take place. The body will adapt to this stimulus. Once the body has adapted, then a different stimulus is required to continue the change. In order for a muscle to increase strength, it must be gradually stresses by working against a load greater than it is used to. Progressive overload not only stimulates muscle hypertrophy, but it also stimulates the development of stronger and denser bones, ligaments, tendons, and cartilage. Progressive overload also incrementally increases blood flow to exercised regions of the body and stimulates more responsive nerve connections between the brain and the muscles involved. If this stress is removed or decreased, there will be a decrease in that particular component of fitness. A normal amount of exercise will maintain the current fitness level. However, if the stress is too much, or increase in stress is not enough progressive, the stress can lead to musculoskeletal injuries.

With the aim to go ahead and to promote sports injury prevention, the methodological sequence for injury prevention research has been described by van Mechelen et al. [68] in four steps: (1) identifying the magnitude of the problem (incidence, characteristics, and severity of sports injuries), (2) determining the risk factors and injury mechanisms, (3) introducing measures that are likely to reduce the future risk and/or severity of sports injuries, and (4) measuring the effectiveness of prevention measures by repeating the first step and/or by means of a randomized clinical trial. This sequence can help scientists to better understand and to improve knowledge on sports injury prevention, but it can also help sports medicine professionals to introduce injury prevention measures among their athletes. Following this sequence, epidemiological studies are thus the first and fundamental step of injury prevention. These are important to better understand the main problems of athletes.

In this context, the aims of this chapter are (1) to present the main methodological aspects of epidemiology in sports injury, (2) to illustrate the epidemiology in sports injuries according to different sports, and (3) try to evaluate the socioeconomic costs of sports-related injury.

1.2 Epidemiological Methodology: Injury Definition and Study Design

The epidemiological data collection represents the first step of the injury prevention sequence [68]. As a first step, it represents a fundamental step. Epidemiological data collection is thus a big challenge for sports medicine staff.

The quality of the data collection methods ensures the quality of epidemiological data. A clear, reproducible and valid method is fundamental to allow comparison of data between studies and to allow long-term follow-up of athletes. The methods of injury surveillance should provide a clear definition of injury (injury, injury receiving medical attention, time-loss injury, subsequent injury, etc.) and its nature (mode of onset, location, type, cause, severity, etc.). The definition of injury precises what injury will be collected during epidemiological studies. Thus, the results of epidemiological studies with different injury definitions cannot be compared. In practice, when you want to analyze injury incidence over different seasons, you have to determine the definition of injury and keep the same throughout the studies. This parameter should also be taken into account when reading epidemiological studies. Moreover, the study design, data collection procedures, and analysis should also be determined and be similar to allow comparison between studies. This includes how to record the data (retrospectively or prospectively and the tool: paper, informatics, Internet, SMS), who records the data (physicians, medical teams, sportsmen, coaches, etc.), what are the exposure (per athlete, per hour of practice, per hour of training, per hour of competition, etc.), and how to calculate prevalence and incidence. Again, differences in injury data collection procedure can lead in differences in the results of injury incidences but without real injury incidence differences. In addition, the methods of injury surveillance should also provide what athlete baseline information is needed (age, height, weight, level of practice, training (volume and intensity), discipline, previous injuries, and/or illnesses) to describe the population and in order to anticipate on the understanding of injury risk factors (step 2 of the injury prevention sequence).

In this context, consensus statements have been developed to standardize epidemiological studies for team sports (e.g., cricket [50], soccer/football [24, 28, 32], rugby union [25, 71], rugby league [38]), for individual sports (e.g., tennis [52], ski [23], horse racing [66], athletics [64]), and for multi-event competitions, such as Olympic Games [34]. Recently, a method has been proposed to prospectively collect overuse injuries, which are often neglected with standard injury surveillance methods [10].

1.3 Epidemiology of Injuries During Olympic Games

Epidemiological data of injuries during Olympic Games (OG) are of interest since it concerns top-level athletes with the same rigorous consensual method [34]. However, these data present a limitation since these are focused on only new injuries and not preexisting injuries and on a short duration (few days or weeks) in comparison to the duration of the whole season.

During the 2004 OG [33], an incidence of 0.8 injuries per match (95 % confidence interval (95%CI), 0.75–0.91) was reported in 14 team sports tournaments (men's and women's football, men's and women's handball, men's and women's basketball, men's and women's field hockey, baseball, softball, men's and women's water polo, and men's and women's volleyball). Half of all injuries affected the lower extremity; 24 % involved the head or neck. The most prevalent diagnoses were head contusion and ankle sprain. Forty-two percent of injuries were expected to prevent the players from sports participation (time-loss injuries). The most common cause was a contact with another player (78 %).

During the 2008 OG [35], injury incidence was 96.1 injuries per 1000 registered athletes, including about 50 % of time-loss injuries, for all disciplines. The majority (72.5 %) of injuries were incurred in competition. One third of the injuries were caused by contact with another athlete, followed by overuse (22 %) and noncontact incidences (20 %). Injuries were reported from all sports, but their incidence and characteristics varied substantially: overall, injury incidences were higher in football (soccer), taekwondo, hockey, handball, weightlifting, and boxing (all > 15 % of the athletes) and lowest for sailing, canoeing/kayaking, rowing, synchronized swimming, diving, fencing, and swimming; and location, types, and causes varied between sports.

During the 2012 OG [21], injury incidence was 128.8 injuries per 1000 registered athletes, including 35 % of time-loss injuries, for all disciplines. The most commonly reported injury mechanisms were overuse (25 %), noncontact trauma (20 %), contact with another athlete (14 %), and contact with a stationary object (12 %). The risk of an athlete being injured was highest in taekwondo, football, BMX, handball, mountain biking, athletics, weightlifting, hockey, and badminton and lowest in archery, canoe slalom and sprint, track cycling, rowing, shooting, and equestrian.

During the 2010 Winter OG [20], injury incidence was 111.8 injuries per 1000 male registered athletes, including 23 % of time-loss injuries and one death. Main locations were the face, head, cervical spine, and the knee. The most common reported injury mechanisms were a noncontact trauma (23.0 %), contact with a stagnant object (21.8 %), and contact with another athlete (14.5 %). The risk of sustaining an injury was highest for bobsleigh, ice hockey, short track, alpine freestyle, and snowboard cross (15–35 % of registered athletes were affected in each sport) and lowest for the Nordic skiing events (biathlon, cross-country skiing, ski jumping, Nordic combined), luge, curling, speed skating, and freestyle moguls (less than 5 % of registered athletes).

During the 2014 Winter OG [62], injury incidence was 140 injuries per 1000 male registered athletes, including 39 % of time-loss injuries. The most commonly reported injury causes/mechanisms were contact with a stationary object (25 %), overuse with gradual onset (14 %), and noncontact trauma (13 %). The injury risk was highest in aerial skiing, snowboard slopestyle, snowboard cross, slopestyle skiing, half-pipe skiing, moguls skiing, alpine skiing, and snowboard half-pipe.

An overview of the injury incidence and characteristics during the Olympic Games is presented in Table 1.1.

1.4 Epidemiology of Injuries in Team Sports

1.4.1 Football

In players of the Union of European Football Association, Ekstrand et al. [19] reported, during a prospective cohort study over 7 seasons using a consensual methodology [28], 8 time-loss injuries per 1000 h of practice among 2226 football players. Incidence was higher during match than training. Most of injuries were located on the thigh (≈20 %) and the knee (≈20–30 %) and involved the tendon (≈20 %) and muscles (≈30 %) [19]. During world football tournaments from 1998 to 2012, using the same methods as Junge et al. [32], Junge and Dvorak [36] reported a total of 3944 injuries from 1546 matches, equivalent to 2.6 injuries per match; about 40 % were time-loss injuries. The majority of injuries (80 %) were caused by contact with another player, compared with 47 % of contact injuries by foul play. The most frequently injured body parts were the ankle (19 %), lower leg (16 %), and head/neck (15 %). Contusions (55 %) were the most common type of injury, followed by sprains (17 %) and strains (10 %).

1.4.2 Rugby

In rugby union, Williams et al. [71] in a meta-analysis reported an overall incidence of injuries in senior men's professional rugby union of 81 injuries per 1000 player hours of match and 3 injuries per 1000 player hours of training, without differences in injury severity between match and training injuries. Muscle/tendon and joint (non-bone)/ligament injuries were the two most prevalent injury groups. In a retrospective analysis of catastrophic cervical spine injuries in French rugby during the period 1996–2006 (including all injuries causing neurological disor-

Table 1.1 Injuries during the last four Olympic Games

	Number of different sports/events	Number of registered athletes	Incidence of injuries per 1000 registered athletes	Percentage of time-loss injuries	Main location or diagnoses	Main causes	Disciplines with higher injury risk
Summer OG 2008 [35]	28/302	10,977	96.1	50	Ankle sprain Thigh strain	Contact with another athlete (33 %), overuse (22 %), noncontact (20 %)	Football (soccer), taekwondo, hockey, handball, weightlifting, and boxing
Summer OG 2012 [21]	26/302	10,568	128.8	35	Thigh, knee, and lumbar spine Sprain, strain, and contusion	Overuse (25 %), noncontact trauma (20 %), contact with another athlete (14 %), contact with a stationary object (12 %)	Taekwondo, football, BMX, handball, mountain biking, athletics, weightlifting, hockey, and badminton
Winter OG 2010 [20]	7/86	2,567	111.8	23	Face, head and cervical spine, and knee Contusion, sprain, strain	Noncontact trauma (23.0 %), contact with a stagnant object (21.8 %), contact with another athlete (14.5 %)	Bobsleigh, ice hockey, short track, alpine freestyle, and snowboard cross
Winter OG 2014 [62]	7/98	2,780	140	39	Knee sprain (first time-loss injury diagnosis)	Contact with a stationary object (25 %), overuse with gradual onset (14 %), noncontact trauma (13 %)	Aerial skiing, snowboard slopestyle, snowboard cross, slopestyle skiing, half-pipe skiing, moguls skiing, alpine skiing, and snowboard half-pipe

OG Olympic Games

der), the rate of injury was 2.1 per 100,000 players per year during the 1996–1997 season and 1.4 during the 2005–2006 season ($p < 0.01$) because of the effectiveness of the preventive measures as the modification of the rules of scrum [6].

1.4.3 Handball

Seil et al. [59] reported, in German handball players, 2.5 time-loss injuries per 1000 h among 186 handball players during one season. Traumatic injuries were located on the knee (10 %), the fingers (9 %), the ankle (8 %), and the shoulder (7 %), and overuse injuries were located on the shoulder (19 %) and the low back (17 %). In a prospective study in a cohort 517 elite Danish handball players, Moller et al. [46] reported 6.3 injuries per 1000 h of practice (training and match) over 31 weeks. 37 % were overuse injuries and 63 % were traumatic injuries. Traumatic injuries were sprains (46 %), muscle strains (17 %), and contusions (9 %) located on the ankle/foot (23 %) and the knee (15 %), and overuse injuries were shin

Table 1.2 Injuries during the international athletics championships

	Incidence of injuries per 1000 registered athletes	Percentage of time-loss injuries	Main diagnosis	Main causes	Events with higher injury risk
World and OG	109	49	Thigh (hamstring) strain	Overuse	Combined events, marathon, middle and long distances
European	65	47	Thigh (hamstring) strain	Overuse	Combined events, middle and long distances
Indoor	69	39	Thigh (hamstring) strain	Overuse and noncontact trauma	Combined events and long distances

Data from 14 athletics championships from 2007 to 2014
OG Olympic Games

splints (22 %), tendinopathy (22 %), and bursitis (7 %) on the knee and the shoulder.

1.4.4 Volleyball

In volleyball, Verhagen et al. [69] reported, in a population of the national Dutch volleyball players, 2.6 time-loss injuries per 1000 h over 1 year. Ankle injuries represented 41 % of all traumatic injuries. Overuse injuries were located on the back (32 %) and the shoulder (32 %).

1.5 Epidemiology of Injuries in Individual Sports

1.5.1 Athletics

In athletics (or track and field), the prevalence of injuries is high (3–170 % per year) [15, 16]. During international athletics championships, data were collected following a consensual similar methodology [1, 34]. Data are quite similar between European and world championships: 10 % of athletes incur an injury, including about 50 % of time-loss injuries [17, 22]. The risk of injury varies substantially between the disciplines with athletes competing in combined events, steeplechase, and middle- and long-distance runs having the highest risk. The most common diagnoses are hamstring strain (16 %), lower leg strain (5–9 %), ankle sprain (3–6 %), and trunk muscle cramps (6 %) [16, 17, 22]. An overview of the injury incidence and characteris-

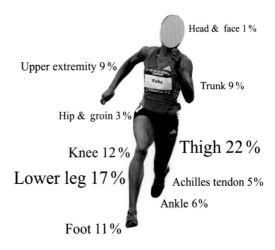

Head & face 1 %

Upper extremity 9 %

Trunk 9 %

Hip & groin 3 %

Knee 12 %

Thigh 22 %

Lower leg 17 %

Achilles tendon 5 %

Ankle 6 %

Foot 11 %

Fig. 1.1 Main injury location for female athletes during international athletics championships from 2007 to 2014 [18]

tics during the international athletics championships is presented in Table 1.2. Main injury location is presented in Fig. 1.1 for female athletes and Fig. 1.2 for male athletes [18]. During the whole season, the current knowledge of injury risk is based on few studies with different methods of injury surveillance, which makes comparison between them and conclusion difficult [17]. An incidence of about 4 injuries per 1000 h of training has been reported in two cohort studies in Australian athletes [3] and Swedish athletes [30]. The characteristics of injuries vary between disciplines, according to the biomechanical and technical movements, implements used, duration of practice, and training workload [15, 16]. Higher acute injury risk is reported in explosive events (sprint, hurdles, and jumps), and higher chronic injury risk is reported in middle- or long-distance runs [5, 15]. In general, most injuries are to the lower limbs (from 60 to 100 %) and involve musculotendinous structures (Fig. 1.3) [15].

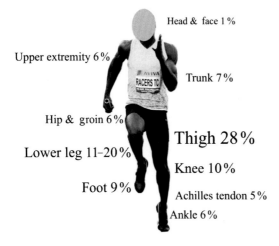

Head & face 1 %

Upper extremity 6 %

Trunk 7 %

Hip & groin 6 %

Thigh 28 %

Lower leg 11–20 %

Knee 10 %

Foot 9 %

Achilles tendon 5 %

Ankle 6 %

Fig. 1.2 Main injury location for male athletes during international athletics championships from 2007 to 2014 [18]

1.5.2 Swimming

During the 2009 FINA World Championships (Aquatics Championships), Mountjoy et al. [47] reported 66 injuries per 1000 registered athletes, including 13 % of time-loss injuries, using the consensual methods from the multi-event competition [38]. Most injuries affected the upper extremity (37 %), followed by the lower extremity (28 %), head/neck (19 %), and trunk (16 %). The most frequently injured body parts were the shoulder (15 %) and head (12 %). Overuse was the most important cause of injuries (38 %). The general injury risk was highest for diving and lowest for swimming. During the 2013 FINA World Championships, Mountjoy et al. [49] reported 83 injuries per 1000 registered athletes. The most common injured body part was the shoulder (21 %). The most common cause of injury was contact with another athlete (25 %). The highest injury incidence rate was in water polo, followed

Fig. 1.3 Summary of the main injury location by sports (traumatic injuries in *red continuous line*; overuse injuries in *orange continuous line*). *Data and percentage have not been reported since these results were not from a system-* *atic exhaustive review and since methods are different between studies making comparison between sports not possible. The aim was to illustrate the main tendencies of sports injuries according to the different sports*

by open water swimming and diving, and the lowest incidence was seen in synchronized swimming.

1.5.3 Skiing

In skiing, the International Ski Federation reported an injury incidence of 36 injuries per 100 athletes per season, in World Cup alpine skiers, including all training sessions and races, over six consecutive seasons, by retrospective interviews [4]. The majority of injuries were time-loss injuries (82 %), leading to the absence from training and competition for at least 1 day. The most common injury location was the knee (38 %), followed by the hand, finger, and thumb (11 %); head/face (10 %); lower leg/Achilles tendon (9 %); lower back, pelvis, and sacrum (9 %); and shoulder/clavicle (7 %). Moreover, in a prospective cohort study in Swedish skiers, Westin et al. [70] reported an injury incidence of 1.6–1.8 time-loss injuries per 1000 ski hours over 5 years. The knee was the most important location (62 %), following by the hand (26 %). For noncompetitive skiers, the risk of injury is about 2.64 per 1000 skier-day (skier day is assimilated to the use of a ticket per day and per skier and is given in France by Domaines Skiables de France (http://www.domaines-skiables.fr)), but the risk is higher (2.85/1000 SD) for snow-boarders. ACL ruptures represent 18 % of injuries and 30 % for women over 15 years old [39].

1.5.4 Cycling

In cycling, De Bernardo et al. [14] reported 112 time-loss injuries (48.5 % traumatic and 51.5 % overuse) among 51 top-level male cyclists, taken part in the main European cycling competitions, in a retrospective interview over the 4 last years. 84 % sustain at least one injury. Traumatic injuries were located on the shoulder girdle (34 %) and overuse injuries on the lower extremity (68 %) [14].

1.5.5 Climbing

Jones et al. [31] reported that around 50 % of climbers had sustained at least one time-loss

injury in the past 12 months, among 201 British rock climbers. Fingers (≈35–30 %) and shoulder (≈20–18 %) were the most frequent location of overuse and acute strenuous injuries. Lefevre and Fleury [40] studied climbing accidents in France during one winter season and found that more than 44 700 had been injured between December 2001 and May 2002.

1.6 Socioeconomic Aspects of Sports-Related Injury

1.6.1 Direct Costs

A general description of sports injuries was published in 2007 by the French Institute for health safety (Institut de veille sanitaire – InVS), based on the 2004–2005 survey on domestic accidents [55]. The data collection is based on the recording of emergency accidents in 12 hospitals in France. Of these, 32 007 (17,8 %) were sports injuries (70,3 % for men and 29,7 % for women) with large disparities by age: 86 % of these injuries occurred for those under 35. More than two accidents in five occurred during sports team practice. This mainly concerned men (83 %) for football practice (70 %), basketball (10 %), rugby (9 %), handball (7 %), and volleyball (3 %). Among women, 31 % of accidents occurred in basketball, 27 % in handball, 21 % in football, and 16 % and 5 % in volleyball and rugby, respectively. Ski and riding accidents were more severe, causing fractures (30 % of lesions) and requiring hospitalization in 16 % and 17 %, respectively. Hospitalizations resulting from ski injuries lasted on average 4.6 days.

In 1995, the French "Sport et Santé" report studied more than 8000 individuals aged 15–49. This sample was divided into three categories depending on whether they practiced sports. The groups were (i) the nonparticipants, (ii) moderate participants (those who practice sports less than 500 h a year), and (iii) very active participants (more than 500 h per year). Geneste et al. [27] showed that the nonparticipants suffered less traumas than the moderates participants (54,8 % versus 66.9 %) and that traumas were five times higher in the very active group compared to the nonparticipants.

The French household survey for health and social protection carried out by IRDES (Enquête Santé et protection sociale)[6] enables a characterization of sports injuries by social profile. According to the 2002 to 2004 surveys, sports accidents involved more often young men aged 10–24 from higher socioeconomic classes [13, 63].

A number of studies have documented the costs of sports (accidents, fractures, sprains, deaths), with a special focus by Garry [26] on sports injuries for the young population.

Sports accidents generate direct costs affecting the victim and the insurer and also indirect costs that are borne by their employer or society at large.

1.6.2   The Benefits Associated with Practicing Sports

A pioneering study in France reported by Marini [44] shows that those practicing sports have an absentee rate of 3.64 % compared to an average rate of 6.58 % for the non practising. The French National Association for the Promotion of Physical Activity and Sports at work (ASMT) has also shown that the frequency of professional accidents decreases with the increase in physical activity.

In Toronto, Cox et al. [12] showed that companies offering physical activity to workers have successfully decreased their absenteeism rate. Van den Bossche [67] confirms that benefits exceed costs for these interventions.

In the United States, some authors have studied worksite health promotion programs and estimated savings at $272 per worker or $26 billion in total. Shephard [60, 61] estimated savings from the health improvements obtained through the practice of sports activities at comparable levels ($211 per individual).

More recently, Pelletier et al. [51] and Schultz and Eddington [58] examined the relationship between changes in health risks and changes in work productivity. Analysis (pre and post) was

conducted on 500 individuals who registered with a wellness program. The results showed that those who reduced their health risk improved their presenteeism by 9 % and reduced absenteeism by 2 %, controlling for risk level, age, and gender.

Mills et al. [45] evaluated the impact of a multi-component workplace health promotion program on employees' health risks and work productivity in the United Kingdom. This program included questionnaires to measure health risks and access to wellness literature, as well as health-related seminars and workshops. The results point to the reduction of health risk factors and absenteeism rates, as well as productivity increases.

Enhancing health through interventions to develop worksite physical activities faces three challenges. On the one hand, such interventions can be perceived by employees as interfering with privacy. On the other hand, these interventions may be more difficult to implement in small- and medium-sized enterprises (SMEs), which will provide an additional source of differences in treatment between employees of large companies and SMEs [53]. Finally, the development of worksite physical activity necessarily implies a change in the organization of work [65, 68].

1.6.3   Special Case of High-Performing Sports

In the special case of high-performing sports, health is put at risk in most cases, meaning that the beneficial effects of sports disappear when practice becomes too intensive. Most of the literature in this area concentrates on the issue of doping activities.

Bourg [8] carried out a cost-benefit analysis to explain drug intake. He showed that there is a net financial benefit derived from taking drugs in case of victory, even after taking into account the psychological consequences associated with the drug intake. However, he acknowledges the fact that doping activities still exist when monetary gains are low (amateur sports events, bodybuilding). Breivik [9] analyzed doping activities using game theory and explained that if victory is the

---

[6]Les accidents de la vie courante en France selon l'Enquête santé et protection sociale 2002. Institut de veille sanitaire, août 2005

most important outcome, then this illegal activity will increase.

In France, the sports ministry has increased funding earmarked to anti-doping campaigns by 4.2 in 6 years, from 5.6 to 24.2 million euros between 1997 and 2002. Maitrot [43] suggests that these anti-doping programs should not only be defined at national level but rather at international level.

Conclusions

Epidemiological data are fundamental and represent the first step of the sequence of prevention. Epidemiological studies are important to improve scientific knowledge on sports injury, to better understand the main injuries, to develop specific research programs, to propose adapted prevention measures focused on the most frequent and/or most severe injuries, to validate prevention measures, and to monitor long-term changes. Epidemiological studies should be based on clear and consensual methods to allow comparison between studies and with time.

The present chapter is not intended to be an exhaustive report of epidemiological data, but it aims illustrating the main injury tendencies between different sports, in order to highlight the interest and need of conducting epidemiological studies to better understand each sports specificity. All sports have different rules, goals, techniques, and tools. Each sport requires specific training and leads to different sportsmen morphologies and to different constraints. Consequently, injuries will be different according to the sports, corresponding to different injury risks, location, and/or types [1, 3]. Injury prevention measures should thus be focused on these main injuries by taking into account the sports constraints.

This chapter opens the door of the next steps of the injury prevention sequence [68]. It illustrates the sports-specific-related injuries for which further investigations should determine the risk factors and mechanisms, which allow injury prevention measures development. Injury prevention measures should be sports specific and be based on the specific injury risk factors and mechanisms. These measures can include different dimensions: athlete's screening and monitoring; improvement of physical condition and sensorimotor control; improvement of sports-specific skills, techniques, and/or biomechanics; optimized medical care of injuries; optimized hygiene of life; and/or education of athletes.

Competing Interest None declared.

References

1. Alonso JM et al (2009) Sports injuries surveillance during the 2007 IAAF World Athletics Championships. Clin J Sport Med 19(1):26–32
2. Andreff W, Szymansky S (2005) Physical activity, sport and health. In: Elgar E (ed) Handbook on the economics of sport. Edward Elgar Publishing, p 143
3. Bennell KL, Crossley K (1996) Musculoskeletal injuries in track and field: incidence, distribution and risk factors. Aust J Sci Med Sport 28(3):69–75
4. Bere T et al (2014) Sex differences in the risk of injury in World Cup alpine skiers: a 6-year cohort study. Br J Sports Med 48(1):36–40
5. Boden BP et al (2012) Catastrophic injuries in pole vaulters: a prospective 9-year follow-up study. Am J Sports Med 40(7):1488–1494
6. Bohu Y et al (2009) Declining incidence of catastrophic cervical spine injuries in French rugby: 1996–2006. Am J Sports Med 37(2):319–323
7. Bouchard C, Shephard RJ, Stephens T (eds) (1994) Physical activity, fitness, and health. International Proceedings and Consensus Statement. Champaign: Human Kinetics
8. Bourg JF (2000) Contribution à une analyse économique du dopage. Reflets et Perspectives de la vie économique 33(2/3):169–178
9. Breivik G (1992) Doping games; a game theoretical exploration of doping. Int Rev Sociol Sport 27(3):235–253
10. Clarsen B, Myklebust G, Bahr R (2013) Development and validation of a new method for the registration of overuse injuries in sports injury epidemiology: the Oslo Sports Trauma Research Centre (OSTRC) overuse injury questionnaire. Br J Sports Med 47(8):495–502. doi:10.1136/bjsports-2012-091524, Epub 2012 Oct 4
11. Conn JM, Annest JL, Gilchrist J (2003) Sports and recreation related injury episodes in the US population, 1997–99. Inj Prev 9(2):117–123
12. Cox M, Shephard RJ, Corey RJ (1981) Influence of an employee fitness programme upon fitness, productivity and absenteeism. Ergonomics 24:795–806
13. Dalichampt M, Thélot B (2008) Les accidents de la vie courante en France métropolitaine – Enquête santé

et protection sociale 2004. Institut de veille sanitaire, Saint- Maurice

14. De Bernardo N et al (2012) Incidence and risk for traumatic and overuse injuries in top-level road cyclists. J Sports Sci 30(10):1047–1053
15. Edouard P et al (2011) Prevention of musculoskeletal injuries in track and field. Review of epidemiological data. Sci Sports 26:307–315
16. Edouard P, Alonso JM (2013) Epidemiology of track and field injuries. New Studies in Athletics 28(1/2):85–92
17. Edouard P, Branco P, Alonso JM (2014) Challenges in Athletics injury and illness prevention: implementing prospective studies by standardised surveillance. Br J Sports Med 48(7):481–482
18. Edouard P et al (2015) Sex differences in injury during top-level international athletics championships: surveillance data from 14 championships between 2007 and 2014. Br J Sports Med 49(7):472–477
19. Ekstrand J, Hagglund M, Walden M (2011) Injury incidence and injury patterns in professional football: the UEFA injury study. Br J Sports Med 45(7): 553–558
20. Engebretsen L et al (2010) Sports injuries and illnesses during the Winter Olympic Games 2010. Br J Sports Med 44(11):772–780
21. Engebretsen L et al (2013) Sports injuries and illnesses during the London Summer Olympic Games 2012. Br J Sports Med 47(7):407–414
22. Feddermann-Demont N et al (2014) Injuries in 13 international Athletics championships between 2007–2012. Br J Sports Med 48(7):513–522
23. Florenes TW et al (2011) Recording injuries among World Cup skiers and snowboarders: a methodological study. Scand J Med Sci Sports 21(2):196–205
24. Fuller CW et al (2006) Consensus statement on injury definitions and data collection procedures in studies of football (soccer) injuries. Br J Sports Med 40(3):193–201
25. Fuller CW et al (2007) Consensus statement on injury definitions and data collection procedures for studies of injuries in rugby union. Br J Sports Med 41(5):328–331
26. Garry F (1999) Les accidents de sport chez 10–24 ans. CNAMTS Point Stat, No. 14
27. Geneste C, Blin P, Nouveau A, Krezentowski R, Chalabi H, Ginesty J, Guezennec Y (1998) Sport et santé: Enquête transversale auprès de 866 sportifs modérés, intensifs ou non sportifs. Santé Publique 1:17–27, In Elgar E (ed) Handbook on the Economics of Sport
28. Hagglund M et al (2005) Methods for epidemiological study of injuries to professional football players: developing the UEFA model. Br J Sports Med 39(6):340–346
29. Inserm(dir) (2008) Activité Physique. Contextes et effets sur la santé. Inserm, Paris. 811 p
30. Jacobsson J et al (2013) Injury patterns in Swedish elite athletics: annual incidence, injury types and risk factors. Br J Sports Med 47(15):941–952
31. Jones G, Asghar A, Llewellyn DJ (2008) The epidemiology of rock-climbing injuries. Br J Sports Med 42(9):773–778
32. Junge A et al (2004) Football injuries during FIFA tournaments and the Olympic Games, 1998–2001: development and implementation of an injury-reporting system. Am J Sports Med 32(1 Suppl):80S–89S
33. Junge A et al (2006) Injuries in team sport tournaments during the 2004 Olympic Games. Am J Sports Med 34(4):565–576
34. Junge A et al (2008) Injury surveillance in multisport events: the International Olympic Committee approach. Br J Sports Med 42(6):413–421
35. Junge A et al (2009) Sports injuries during the Summer Olympic Games 2008. Am J Sports Med 37(11):2165–2172
36. Junge A, Dvorak J (2013) Injury surveillance in the World Football Tournaments 1998–2012. Br J Sports Med 47(12):782–788
37. Kamitani T et al (2013) Catastrophic head and neck injuries in judo players in Japan from 2003 to 2010. Am J Sports Med 41(8):1915–1921
38. King DA et al (2009) Epidemiological studies of injuries in rugby league: suggestions for definitions, data collection and reporting methods. J Sci Med Sport 12(1):12–19
39. Laporte JD (2013) L'accidentologie des sports d'hiver. Saison 2012–2013. Dossier de presse. Association des Médecins de Montagne. Available via AMM. http://www.mdem.org. Accessed 3 Avril 2015
40. Lefevre B, Fleury B (2002) Hiver 2001–2002: bilan des interventions. Système National d'Observation de la Sécurité en Montagne, Chamonix
41. Ljungqvist A (2008) Sports injury prevention: a key mandate for the IOC. Br J Sports Med 42(6):391
42. Ljungqvist A et al (2009) The International Olympic Committee (IOC) Consensus Statement on periodic health evaluation of elite athletes March 2009. Br J Sports Med 43(9):631–643
43. Maitrot E (ed) (2003) Le scandale du sport contaminé. Flammarion, Paris
44. Marini F (1980) Sport et travail: Incidence professionnelle de la pratique des sports en compétition. Mémoire CES de biologie et médecine du sport, University of Besançon
45. Mills PR, Kessler RC, Cooper J, Sullivan S (2007) Impact of a health promotion program on employee health risks and work productivity. Am J Health Promot 22(1):45–53
46. Moller M et al (2012) Injury risk in Danish youth and senior elite handball using a new SMS text messages approach. Br J Sports Med 46(7):531–537
47. Mountjoy M et al (2010) Sports injuries and illnesses in the 2009 FINA World Championships (Aquatics). Br J Sports Med 44(7):522–527
48. Mountjoy M, Junge A (2013) The role of International Sport Federations in the protection of the athlete's health and promotion of sport for health of the general population. Br J Sports Med 47(16): 1023–1027

49. Mountjoy M et al (2015) Competing with injuries: injuries prior to and during the 15th FINA World Championships 2013 (aquatics). Br J Sports Med 49:37–43

50. Orchard JW et al (2005) Methods for injury surveillance in international cricket. Br J Sports Med 39(4), e22

51. Pelletier B, Boles M, Lynch W (2004) Change in health risks and work productivity over time. J Occup Environ Med 46(7):746–754

52. Pluim BM et al (2009) Consensus statement on epidemiological studies of medical conditions in tennis, April 2009. Br J Sports Med 43(12):893–897

53. Proper K, Van Mechelen W (2004) Costs, benefits and effectiveness of worksite physical activity counseling from the employer's perspective. Scand J Work Environ Health 30(1):36–46. doi:10.5271/sjweh.763

54. Rees JD, Stride M, Scott A (2014) Tendons-time to revisit inflammation. Br J Sports Med 48(21):1553–1557. doi:10.1136/bjsports-2012-091957, Epub 2013 Mar 9

55. Ricard C, Rigou A, Thélot B (2008) Description et incidence des recours aux urgences pour accidents de sport en France. Enquête permanente sur les accidents de la vie courante, 2004–2005. Bull Epidemiol Hebd 33:293–295, Rapport InVS, décembre 2007. Sur http://www.invs.sante.fr

56. Rigou A (2013) Une estimation des décès traumatiques liés à la pratique sportive en France métropolitaine, en 2010. Journal de traumatologie du sport 30:3

57. Rogge J (2009) An ounce of prevention? Br J Sports Med 43(9):627

58. Schultz AB, Eddington DW (2007) Employee health and presenteeism; A systematic review. J Occup Rehabil 17(3):547–579. doi:10.1007/s10926-007-9096-x

59. Seil R et al (1998) Sports injuries in team handball. A one-year prospective study of sixteen men's senior teams of a superior nonprofessional level. Am J Sports Med 26(5):681–687

60. Shephard RJ (1986) Economic benefits of enhanced fitness. Human Kinetics Publishers, Inc., Champaign, x + 120 p

61. Shephard RJ (1992) A critical analysis of work-site fitness programs and their postulated economic benefits. Med Sci Sports Exerc 24(3):354–370

62. Soligard T et al (2015) Sports injuries and illnesses in the Sochi 2014 Olympic Winter Games. Br J Sports Med 49(7):441–447

63. Thélot B, Ricard C (2005) Résultats de l'Enquête permanente sur les accidents de la vie courante, années 2002–2003. Réseau Epac. Institut de veille sanitaire, Département maladies chroniques et traumatismes, octobre 2005

64. Timpka T et al (2014) Injury and illness definitions and data collection procedures for use in epidemiological studies in Athletics (track and field): Consensus statement. Br J Sports Med 48(7):483–490

65. Toussaint J.-F. (coord.) (2008) Plan national de prévention par l'activité physique ou sportive (PNAPS)

66. Turner M et al (2012) European consensus on epidemiological studies of injuries in the thoroughbred horse racing industry. Br J Sports Med 46(10): 704–708

67. Van den Bossche F (1991) L'impact positif de la généralisation d'une pratique sportive sur les coûts de la sécurité sociale. Paper presented at the conference on Sport, Economy and Politics, Université libre de Bruxelles, 14 Nov 1991

68. van Mechelen W, Hlobil H, Kemper HC (1992) Incidence, severity, aetiology and prevention of sports injuries. A review of concepts. Sports Med 14(2): 82–99

69. Verhagen EA et al (2004) A one season prospective cohort study of volleyball injuries. Br J Sports Med 38(4):477–481

70. Westin M, Alricsson M, Werner S (2012) Injury profile of competitive alpine skiers: a five-year cohort study. Knee Surg Sports Traumatol Arthrosc 20(6):1175–1181

71. Williams S, Trewartha G, Kemp S, Stokes K (2013) A meta-analysis of injuries in senior men's professional Rugby Union. Sports Med 43(10):1043–1055. doi:10.1007/s40279-013-0078-1

Major Causes of Sports Injuries

2

Nicolas Lefevre, Yoann Bohu, Serge Herman, Shahnaz Klouche, and Elvire Servien

Key Points

1. Sports injuries depend on numerous factors, intrinsic (biological characteristics, anatomical factors, sex, age, and stage of development in adolescents) and extrinsic (the practice of sports itself and the environment it is practiced in).

2. Male athletes seem to have a higher risk of severe sports-related injuries, while women are more affected by overuse than men.

3. To prevent secondary injuries, athletes should receive high-quality training and the correct amount of training and recovery and have a healthy lifestyle.

4. Good training equipment is essential to prevent sports-related accidents and overuse injury.

5. Exhaustion and overtraining must be avoided.

N. Lefevre, MD (✉) • Y. Bohu, MD
S. Herman, MD • S. Klouche
Department of Orthopaedic Surgery,
Clinique du Sport Paris, Paris 75005, France

Department of Orthopaedic Surgery,
Institut de l'Appareil Locomoteur Nollet,
Paris 75017, France
e-mail: docteurlefevre@sfr.fr

E. Servien
Department of Orthopaedic Surgery, Hopital de la
Croix-Rousse, Centre Albert Trillat, Lyon University,
Lyon 69004, France

2.1 Overview of Sports Injury

In the past 40 years, there has been a significant increase in the number of men and even more of women who practice sports of all types and at all levels in Europe, the world, and in particular in the United States after amendment n° IX was passed in 1972 on Education [55] against the discrimination of women practicing sports in school. The number of young men practicing sports in the United States in high school has increased by 3 % (from 3.7 to 3.8 million) since 1972, while the number of women has increased ninefold and has doubled every 10 years (from 0.3 to 2.8 million) [43]. This increase is associated with greater exposure to accidents and thus to a risk of injury. Today, both leisure and competitive sports have become an integral part of daily life in Western countries [40, 45]. Physical activity is beneficial to one's health. Medical professionals estimate that regular physical activity significantly decreases the frequency of chronic diseases [3] such as high blood pressure, heart disease [38], colon cancer [20], and diabetes [11] as well as reducing cardiovascular-related deaths [38]. Physical activity is also a remedy for psychological disorders and markedly limits the severity of episodes of anxiety and depression. The counterpart to these health benefits is that daily physical activity is associated with a higher risk of injury [57]. There is a risk of accidents or injury [12] to the musculoskeletal system including soft tissue damage, fractures, ligament and tendon tears, and

H.O. Mayr, S. Zaffagnini (eds.), *Prevention of Injuries and Overuse in Sports: Directory for Physicians,
Physiotherapists, Sport Scientists and Coaches*, DOI 10.1007/978-3-662-47706-9_2

nerve injuries. These may occur in athletes of all ages [39]. Sports-related injuries and incapacitation generally occur in the joints: the knee, ankle, hip, shoulder, elbow, wrist, and spine. Injuries may occur following an acute episode from an accident but also due to overuse, from repetitive microtraumas that are individually insufficient to produce macroscopic injuries [9, 45]. Microtraumatic injury that results when an anatomical structure is exposed to a repetitive, cumulative force where the body's reparative efforts are exceeded and local tissue breakdown occurs is different from overload. The principle of overload states that a greater than normal stress or load on the body is required for training adaptation to take place. The body will adapt to this stimulus. Once the body has adapted, then a different stimulus is required to continue the change. In order for a muscle (including the heart) to increase strength, it must be gradually stressed by working against a load greater than it is used to.

The type of injury therefore depends on numerous interdependent factors, which can be divided into two main causes: intrinsic and extrinsic (Fig. 2.1). Extrinsic factors are linked to the practice of sports itself and the environment it is practiced in. Intrinsic factors include biological characteristics, anatomical factors, sex, age, and stage of development in adolescents, which are going to favor the practice of one sports rather than another [39]. Age is one of the most well-known factors, in particular at the two ends of the spectrum. Thus as the person ages, injuries such as musculotendinous tears, tendinitis, and osteoarthritis become increasingly frequent when sports are practiced frequently. On the other hand during childhood and adolescence, the risk of musculoskeletal injury is high. Indeed, the growth plate has not closed, and accidental injury to growing bones (knee, ankle, back, etc.) is frequent and potentially serious. Athletes are highly motivated to be successful in particular during adolescence and young adulthood, which are periods of physical development. The muscles, bones, and tendons are still immature during this period, and sports injuries are frequent. An analysis of the causes of these accidents requires a good understanding of the accidental process and

sufficient information on those involved in this process (sports doctor, trainer, teacher, coach). Moreover, this analysis should be based on the local practice of sports in situ, for each specific sport (Fig. 2.2). Indeed, the risks are not the same depending on the place because of the geography and climate and local sports traditions (baseball in the United States, rugby in France [6, 7] or the United Kingdom, hockey in Canada) and equipment (scooter, roller skates, snowboards). The risks also change over time: equipment and materials change over time and sports appear and disappear.

Through the decades and the centuries, the world of sports has evolved very differently from one sport to another. Football (invented by the British at the end of the nineteenth century) is the most frequently practiced sport in the world, and the World Cup is the most prestigious international competition. It has been held every 4 years since 1930. At the same time, street football, which was originally practiced in parking lots or abandoned lots, has grown into a more structured ball sport such as futsal, which has five players. It is played in a closed room, on synthetic turf, with a short playing time and with different types of accidents than outdoor football; the first world championships were held in 1982. Skiing has existed since the nineteenth century, and it became an Olympic sport in 1936. Later the monoski was invented in 1961 by Jacques Marchand (several years before the snowboard), but it never became an Olympic sport. After being relatively popular in the 1970s and 1980s, it has practically disappeared today. On the other hand snowboarding became an Olympic sport in 1990 with a very different traumatology profile than traditional skiing. The first roller skates had two wheels under the ball of the foot and two under the heel, "quads," whose shape made it possible to take curves more easily. With the invention of polyurethane wheels in 1979, roller blades were invented with four wheels in a line, and today quads have nearly completely disappeared.

Surfing was invented in Hawaii in the eighteenth century. It became popular throughout the world in the 1950s and 1960s with the first championships in 1976. Kite surf (originally fly surf)

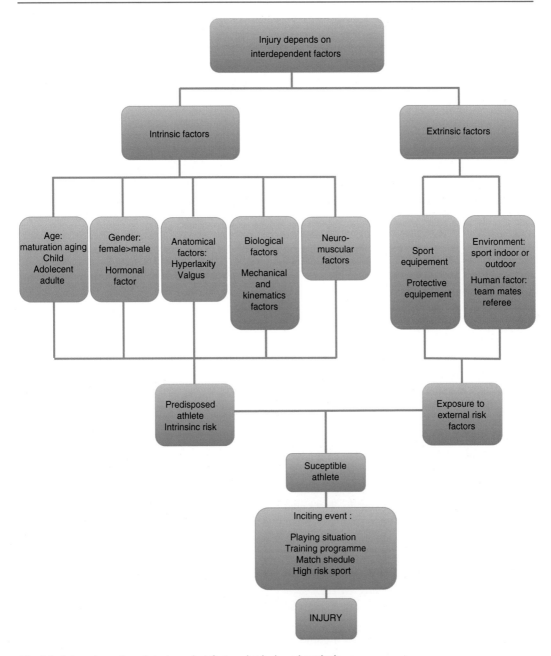

Fig. 2.1   Injury depends on interdependent factors: intrinsic and extrinsic

was imagined by several inventors in the 1960s. After working to improve the sail, two brothers from Quimper, Dominique and Bruno Legaignoux, filed the patent for the curved wing and the inflatable structure in 1984. The first world championships took place in 2000. Kite surfing, with a surfboard and a sail, is a very dan-gerous and even deadly sport. There are very strict rules about practicing kite surfing on cer-tain beaches.

An analysis of sports injury must be adapted to each sport at a particular period. To reduce their numbers it therefore seems logical to try to under-stand how these accident injuries happen. It is

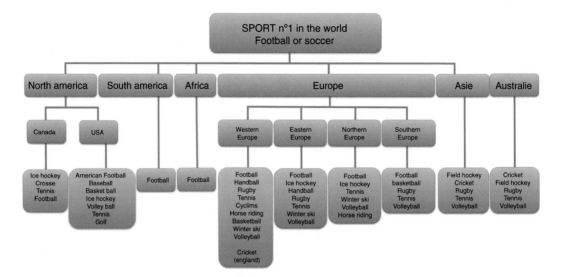

Fig. 2.2 Distribution of sports practiced in the world

also important to take into account all types of athletes, that is, those practicing leisure sports as well as competitive athletes. Indeed, only 10 % of the athletes practice competitive or high-level sports. Thus here when we speak of sports, we mean sports in the broad sense of the term, including a professional football match and occasional leisure games. Finally it is important to analyze the different sports separately: the martial arts include judo [36] karate, jujitsu, Thai boxing, kickboxing, aikido, and tae kwon do. Track and field includes running, jogging, and all other categories of track and field, which makes a relatively heterogeneous group of sports. Classical or modern dance, break dancing, etc., or any other activity associated with dancing, whether it is practiced in a club or at home. Football includes any accidents that occur on the playing field (training or competition) but also in a game on the street. The field of gymnastics and physical exercises are associated with a greater risk of chronic injuries from overuse. Competitive or amateur cyclists have an increased risk of accidents on the road or during the practice of mountain bike. Finally winter sports such as skiing, snowboarding, or hockey are more often associated with severe accidents and injuries. Sports injuries must also be analyzed over time: several studies have shown that there is a certain rhythm

to accidents in adults as well as in children. A decrease in the number of accidents during school vacations has been observed, mainly summer vacation but also to a lesser extent in spring, autumn, and winter. Indeed, not only do children practice most sports during gym at school, but many sports clubs also take a break during school vacations. Two peaks can also be noted for injuries in numerous sports: a peak in the spring because of children's physiological hyperactivity during this period and in the fall due to fatigue, which is also physiological. The accident curves of certain sports are more season dependent such as skiing or ice-skating in the winter and swimming, tennis, and track and field in the summer.

2.2 Physical Constitution

A clear increase in the number of accidents can be noted for nearly all sports during early adolescence. This increase is of course because children of that age practice more sports but also because they take greater risks, especially boys, by practicing sports that are less well supervised than during childhood. It can be noted that the increase in the number of accidents during adolescence happens earlier and more suddenly in school sports especially (gymnastics and ball sports) which, by definition, are practiced by both boys and girls equally. This increase in injuries is followed by a peak then a decrease. In comparing the sexes and all sports combined, this peak occurs at 14 years old for boys and 12 years old for girls depending on the series [10, 31, 49, 53]. It corresponds to the pubescent growth spurt that occurs between 12 and 13 for girls and approximately a year or a year and a half later for boys. The puberty results in profound physical and psychological changes. These changes require major psychomotor readaptation and result in irregular physical performance, which can include a certain clumsiness, thus a greater risk of accidents. For Narring et al. [42] this is especially obvious at puberty because the morphological and psychological developmental changes of girls and boys differ. A lack of practice of motor skills during a period when the body needs to adapt to locomotor changes is certainly a handicap for girls and can increase their risk of accidents. Moreover, body mass increases, especially fat, while boys develop muscle. Thus girls have an unfavorable weight/strength ratio, which can create problems.

2.3 Sex Differences

There are important differences between the sexes in the causes of injury. There are anatomical, physiological, and psychological criteria [8, 45] as well as differences according to the type of sports practiced. Male athletes seem to have a higher risk of severe sports-related injuries [47], while women are more affected by overuse than men [14, 26, 50, 61]. Distribution of sports injuries between the sexes in the literature is on average 60 % for boys and 40 % for girls, all sports included [15, 31, 51, 56, 58]. Certain studies reported a less marked difference with 51–55 % of injuries for boys [4, 30, 36, 44], while in other series there was a greater difference with 71 % of male injuries [10, 54]. This male predominance is logical if each sport is studied separately. It is most marked in football, where girls only represent 6.5 % of the cases. The only sports where the male predominance is reversed are volleyball, handball, gymnastics, horseback riding, and figure skating [5, 13, 22, 23,

27, 49, 51, 58]. Most of the time this male or female predominance can simply be explained by a greater participation of boys than girls in a specific sport: football and ice hockey are mainly male sports, while dance, skating, and horseback riding are more frequently practiced by girls. All sports combined, the types of injuries are basically the same for the two sexes even if there are some slight variations. The main injury in girls is sprains with a mean 21.5–44 % of sprains versus 16–24 % in boys [15, 29]. Hormonal factors are one of the most important risk factors explaining the difference between the sexes. Numerous studies have evaluated the risk factors of anterior cruciate ligament (ACL) tears in women (Table 2.1) and noted hormonal factors. We know that there are hormone receptors on the ACL in particular estrogen, progesterone, testosterone, and relaxin receptors. Although it is known that these receptors affect the metabolism and mechanical properties of the ACL, at present we do not know what the actual effects are [16, 18, 25, 37]. A review of the literature has shown that the frequency of ACL injuries is not constant during the menstrual cycle. There is a significantly greater frequency of tears during the preovulatory phase than during the postovulatory phase. The study by Wojtys clearly shows a greater prevalence of ACL tears (noncontact pivot) in female athletes during the preovulatory phase of the menstrual cycle [60], and other studies reach the same conclusion [1, 34, 52]. Severe sprains during the preovulatory phase were significantly more frequent than during the postovulatory phase (OR, 3.22). This distinction can be found in numerous epidemiological studies in particular the study by Dugan in 2005 [17], which showed that women athletes who practice sports involving jumps and intermittent efforts had a risk of knee injury that was two to six times greater than their male counterparts.

2.4 Training Deficits

A significant increase in the intensity of training and changes in the type of activity in sports, and in particular in sports techniques are considered to be risk factors for injuries in the presence of

Table 2.1 Risk factors of anterior cruciate ligament (ACL) tears in women

Study	Sport	Female incidence (per 100,000)	Male incidence (per 100,000)	Sex ratio
Myklebust et al. (1998) [41]	Handball	82	31	2.65
Prodromos et al. (2007) [46]	Basket ball	29	8	3.6
	Football	32	12	2.77
	Handball	56	11	5.1
	Fighting sport	77	19	4.01
Parkkari et al. (2008) [44]	All sport	30	96	3.32
Study	Sport	Female incidence (per 1000 h)	Male incidence (per 1000 h)	Sex ratio
Faude et al. (2005) [19]	Football	0 (training) 2.2 (match)	? (training) 1.0 (match)	
Fuller et al. (2007) [21]	Football	0.09 (training) 1.64 (match)	0.03 (training) 0.47 (match)	3.49
Le Gall et al. (2008) [33]	Football	0.02 (training) 1.1 (match)		
Hagglund et al. (2009) [24]	Football	0.15	0.11	1.31
Walden et al (2010) [59]	Football	0.04 (training) 0.72 (match)	0.03 (training) 0.28 (match)	2.26
Rueld et al. (2011) [48]	Ski			2.6

deficient training. To prevent secondary injuries, athletes should receive high-quality training as well as the correct amount of training and recovery and have a healthy lifestyle. Training must absolutely follow strict criteria for regularity, progressivity, adaptation, and recovery. Training should always be preceded by a more or less lengthy warm-up period that is adapted to the specific sport. Warming-up prepares the body for the cardiovascular, pulmonary, muscular, and even psychological stresses to follow. The muscles, organs, and joints will receive better provisions of oxygen and nutrients when the real physical effort begins. To be effective, the warm-up process should follow certain basic rules: it should be long enough. If it is too short, the muscles will tire more quickly. It should be gradual: the effort should be gradually increased and performed faster until it reaches 50–60 % of the athlete's total capacity. It should be adapted to the situation: the warm-up should be associated with general stretching and relaxation to prepare the muscles and joints for stresses and strains. This should be followed by movements specific to each sport. Increasing muscle strength is an integral part of training, to improve performance, but also to prevent accidents. Tonic and balanced muscle tone helps regulate the stresses and strains on tendons, bones, and joints during daily practice as well as under extreme or dangerous

circumstances. Increasing the amount of training, especially suddenly, is associated with a corresponding increase in the number of accidental injuries. Thus for runners, the risk of injury increases when they run more than 35 to 40 kms per week. In the same way, so-called interval training, which alternates several acceleration phases with more or less long recovery phases, is the cause of musculotendinous accidents. Recovery is essential because it allows the body to gradually return to a state of rest. Unfortunately this phase is often too short or even nonexistent. There are different ways to recover: reduce the intensity of the exercises and the frequency of training or simply to rest. Finally a healthy lifestyle is essential for an athlete. Hydration that is regular, continuous, and slightly alkaline, taken before, during, and after training so that the athlete never reaches the threshold of thirst, which is frequently too late, is an effective way to prevent numerous cases of tendinitis. Also the phases of recovery following an effort and in particular the quality of sleep must absolutely be preserved to avoid overtraining. The sports doctor can also suggest a sport to a future athlete that she/he may not have considered but that could be better adapted to his/her morphology or physical capacities. This element is important for the person to be motivated by his/her results and enjoy practicing sports under optimal conditions.

2.5   Inadequate Training Techniques

The sports technique must be perfect; otherwise, injuries can develop: tennis elbow in amateur tennis players who practice a backhand with a non-stabilized wrist which overuses the epicondylar muscles [28] and injury to the posterior cubital tendon from repetitive topspin forehands causing excess tension in this tendon. Good quality equipment should be used that is adapted to the player (morphology and level of play). In tennis, to prevent tennis elbow a lighter racket should be used that is not strung too tightly (less than 23 kgs), with a handle that is not too long without forward balance to avoid overusing the elbow muscles. Good training equipment is essential to prevent sports-related accidents and overuse injury, called technopathies. For example, shoes that are adapted to the sport being played, to the terrain, as well as to the player's weight and feet significantly reduce the risk of developing musculoskeletal problems. In the same way, a bicycle should be adapted and adjusted to the user's morphology and age. Once again, the best advice can be given by trainers or sports doctors who have in-depth knowledge of the sport and its constraints. Based on their experience, athletes, even amateur athletes, can choose the material that is best adapted to them. For certain sports such as cycling, the equipment can be adapted by the morphology of each individual athlete by a specialist, an ergonomist.

The surface of the playing field can also be a source of injuries if the equipment is not adapted: increasing the friction coefficient between sports shoes and the playing surface to improve power and performance also increases the risk of ACL injuries. Lambson reported that the risk of ACL injuries was greater in football players with more cleats and when the playing field was more adherent [32]. Olsen reported a higher risk of ACL in women's handball teams who played on courts with artificial surfaces (with greater adherence) than on wooden courts. This relationship was not found in male athletes [2]. A certain number of factors that favor sports injuries can be identified such as insufficient physical training, poor under- standing of the sports technique, dangerous or inappropriate playing field, poorly adapted equipment or equipment that the player does not know how to use properly, poor training of coaches, inadequate or insufficiently strict rules, lack of surveillance, poor weather conditions, etc. These factors are more or less important depending on the sport. For sports such as vol- leyball or basketball where an impact between the ball and the hand is the cause of most injuries, technique is going to be most important. Group sports and gymnastics, which are the source of 43 % and 26 % of all accidental injuries, are the sports with the most frequent accidents. Nevertheless there is a difference between the sexes: boys are injured more while playing rugby (24 % of males) and girls while doing gymnastics (29 % of women). In more than half these acci- dents, girls hurt themselves alone (due to a poor landing or placement) (Fig. 2.3). In 45 % of the injuries in boys, a third party is involved (impact or foul play by an opponent). In rugby 73 % of the accidents are associated with foul play (e.g., tackling or slightly high tackle) or direct impact with an opponent, while in gymnastics 54 % of the accidents are due to a poor landing.

2.6   Excessive Demands in Training and Competition

Sports injuries are often a sign that the organism can no longer adapt to the demands of the sport. They can easily occur when the intensity or quality of training is suddenly changed, when equipment is changed, or when the sports technique or the way the competitive sport is practiced is changed. To identify this situation the practitioner must have a good understanding of the particularities of each sport. Exhaustion and overtraining must be avoided. This is true for both occasional players and professional athletes. Athletes who do not take into account their limits and their body's warning signs are at risk of injury. When a person is inactive, a week of skiing or intense tennis playing often requires an effort that the body is not prepared for. For example, ski accidents often occur on the third

Fig. 2.3 (**a, b**) Bad pelvic
balance and a poor landing or
placement

day, in the late afternoon. Paradoxically, this same type of problem is also a risk for athletes who overtrain. Increasing training does not necessarily improve performance. Athletes who do not take into account their limits will see their performance worsen. Refusing to accept reality can affect the athlete's physical and mental health. Overtraining prevents the body from recovering. Chronic fatigue sets in increasing the risk of accident-related injuries of all kinds. At the same time, if the athlete begins to perform poorly, this affects his/her motivation and attention and can even cause depression. Studying the mechanisms of sports injuries as well as their consequences and the factors that increase their risk is essential for a better understanding of the relationship between sports and accidents. This highly informative evaluation can help identify the best practices to reduce the frequency of accidents. This information can help precisely determine the conditions and main characteristics of accidents to increase the efficacy of proposed models of prevention.

Two types of variables provide information on the circumstances involved in an injury: the etiopathogenesis and the activity causing the accident. The etiopathogenesis corresponds to the accidental behavior that caused the injury. The activity provides information about the characteristics of sport (football, ski, etc.) that is the cause of the injured limb (upper, lower, right, or left limb), topography (part of the body that is affected), type of injury, sport being practiced, and the time of day the accident occurred. For example, a proximal hamstring tendon tear can occur in cases of excess traction of the tendons during waterskiing [35].

Very little is known about the risk of injury during competitive matches. Myklebust has reported that men or women athletes have a higher risk of an ACL injury during a handball match than during training sessions [41]. Johnson reported a greater number of women with tears during ski competitions with a sex ratio F/M 2.5 [22]. But there are no studies to explain this extrinsic risk factor. There are no studies that have evaluated factors specific to the sports themselves, for example, the rules, the referees, or the trainers. Other factors could also play a role such as the age, level of competence, and psychological profile of the athlete. Rapid changes in the quality and quantity of training or intensive competition are also risk factors for injury [45].

As mentioned previously, the sports with the greatest number of repetitive actions have a higher risk of recurrent injuries; combined with poor technique, the risk increases even more.

References

1. Arendt EA (2007) Musculoskeletal injuries of the knee: are females at greater risk? Minn Med 90:38–40
2. Arnason A, Gudmundsson A, Dahl HA, Johannsson E (1996) Soccer injuries in Iceland. Scand J Med Sci Sports 6:40–45
3. Atay E, Hekim M (2013) The effect of physical activity on health in adults individuals. SSBT Int Ref Acad J Sports 7:113–122
4. Backx FJ, Erich WB, Kemper AB, Verbeek AL (1989) Sports injuries in school-aged children, an epidemiologic study. Am J Sports Med 17:234–240
5. Bixby-Hammett D (1992) Pediatric equestrian injuries. Pediatrics 89:1173–1176
6. Bohu Y, Julia M, Bagate C, Peyrin JC, Colonna JP, Thoreux P et al (2009) Declining incidence of catastrophic cervical spine injuries in French rugby: 1996–2006. Am J Sports Med 37:319–323
7. Bohu Y, Klouche S, Lefevre N, Peyrin JC, Dusfour B, Hager JP et al (2014) The epidemiology of 1345 shoulder dislocations and subluxations in French Rugby Union players: a five-season prospective study from 2008 to 2013. Br J Sports Med. doi:10.1136/bjsports-2014-093718
8. Bohu Y, Klouche S, Lefevre N, Webster K, Herman S (2014) Translation, cross-cultural adaptation and validation of the French version of the Anterior Cruciate Ligament-Return to Sport after Injury (ACL-RSI) scale. Knee Surg Sports Traumatol Arthrosc. doi:10.1007/s00167-014-2942-4
9. Bohu Y, Lefevre N, Bauer T, Laffenetre O, Herman S, Thaunat M et al (2009) Surgical treatment of Achilles tendinopathies in athletes. Multicenter retrospective series of open surgery and endoscopic techniques. Orthop Traumatol Surg Res 95:S72–S77
10. Boyce SH, Quigley MA (2003) An audit of sports injuries in children attending an Accident & Emergency department. Scott Med J 48:88–90
11. Bull FC, Armstrong TP, Dixon T, Ham S, Neiman A, Pratt M (2004) Physical inactivity. In: Ezzati M, Lopez AD, Rodgers A, Murray CJL (eds) Comparative quantification of health risks: global and regional burden

of disease attributable to selected major risk factors. World Health Organization, Geneva, p 729–881

12. Caine DJ, Maffulli N (2005) Epidemiology of children's individual sports injuries. An important area of medicine and sport science research. Med Sport Sci 48:1–7

13. Campbell-Hewson GL, Robinson SM, Egleston CV (1999) Equestrian injuries in the paediatric age group: a two centre study. Eur J Emerg Med 6:37–40

14. Cuff S, Loud K, O'Riordan MA (2010) Overuse injuries in high school athletes. Clin Pediatr (Phila) 49:731–736

15. Damore DT, Metzl JD, Ramundo M, Pan S, Van Amerongen R (2003) Patterns in childhood sports injury. Pediatr Emerg Care 19:65–67

16. Dragoo JL, Lee RS, Benhaim P, Finerman GA, Hame SL (2003) Relaxin receptors in the human female anterior cruciate ligament. Am J Sports Med 31:577–584

17. Dugan SA (2005) Sports-related knee injuries in female athletes: what gives? Am J Phys Med Rehabil 84:122–130

18. Faryniarz DA, Bhargave AM, Lajam C, Attia ET, Hannafin JA (2006) Quantitation of estrogen receptors and relaxin binding in human anterior cruciate ligament fibroblasts. In Vitro Cell Dev Biol Anim 42:176–181

19. Faude O, Junge A, Kindermann W, Dvorak J (2005) Injuries in female soccer players. A prospective study in the German national league. Am J Sports Med 33(11):1694–1700

20. Friedenreich CM, Neilson HK, Lynch BM (2010) State of the epidemiological evidence on physical activity and cancer prevention. Eur J Cancer 46:2593–2604

21. Fuller CW, Dick RW, Corlette J, Schmalz R (2007) Comparison of the incidence, nature and cause of injuries sustained on grass and new generation artificial turf by male and female football players. Part 1: match injuries. Br J Sports Med 41(Suppl 1):i20–i26

22. Fuller CW, Dick RW, Corlette J, Schmalz R (2007) Comparison of the incidence, nature and cause of injuries sustained on grass and new generation artificial turf by male and female football players. Part 2: training injuries. Br J Sports Med 41:i27–i32

23. Giebel G, Braun K, Mittelmeier W (1993) Pferdesportunfälle bei Kindern. Chirurg 64:938–947

24. Hagglund M, Waldén M, Ekstrand J (2009) Injuries among male and female elite football players. Scand J Med Sci Sports 19(6):819–827

25. Hamlet WP, Liu SH, Panossian V, Finerman GA (1997) Primary immunolocalization of androgen target cells in the human anterior cruciate ligament. J Orthop Res 15:657–663

26. Hart DA, Archambault JM, Kydd A, Reno C, Frank CB, Herzog W (1998) Gender and neurogenic variables in tendon biology and repetitive motion disorders. Clin Orthop Relat Res 351:44–56

27. Holland AJ, Roy GT, Goh V, Ross FI, Keneally JP, Cass DT (2001) Horse-related injuries in children. Med J Aust 175:609–612

28. Kannus P (1997) Etiology and pathophysiology of chronic tendon disorders in sports. Scand J Med Sci Sports 7:78–85

29. Kelm J, Ahlhelm F, Anagnostakos K, Pitsch W, Schmitt E, Regitz T et al (2004) Gender-specific differences in school sports injuries. Sportverletz Sportschaden 18:179–184

30. Kelm J, Ahlhelm F, Pape D, Pitsch W, Engel C (2001) School sports accidents: analysis of causes, modes, and frequencies. J Pediatr Orthop 21:165–168

31. Kvist M, Kujala UM, Heinonen OJ, Vuori IV, Aho AJ, Pajulo O et al (1989) Sports-related injuries in children. Int J Sports Med 10:81–86

32. Lambson RB, Barnhill BS, Higgins RW (1996) Football cleat design and its effect on anterior cruciate ligament injuries: a three-year prospective study. Am J Sports Med 24:155–159

33. Le Gall F, Carling C, Reilly T (2008) Injuries in young elite female soccer players. An 8- season prospective study. Am J Sports Med 36(2):276–284

34. Lefevre N, Bohu Y, Klouche S, Lecocq J, Herman S (2013) Anterior cruciate ligament tear during the menstrual cycle in female recreational skiers. Orthop Traumatol Surg Res 99:571–575

35. Lefevre N, Bohu Y, Naouri JF, Klouche S, Herman S (2013) Returning to sports after surgical repair of acute proximal hamstring ruptures. Knee Surg Sports Traumatol Arthrosc 21:534–539

36. Lefevre N, Rousseau D, Bohu Y, Klouche S, Herman S (2013) Return to judo after joint replacement. Knee Surg Sports Traumatol Arthrosc 21:2889–2894

37. Liu SH, Al-Shaikh RA, Panossian V, Yang RS, Nelson SD, Soleiman N et al (1996) Primary immunolocalization of estrogen and progesterone target cells in the human anterior cruciate ligament. J Orthop Res 14:526–533

38. Löllgen H, Löllgen D (2004) Physical activity and prevention of disease. Dtsch Med Wochenschr 129:1055–1056

39. Maffulli N, Longo UG, Gougoulias N, Caine D, Denaro V (2011) Sport injuries: a review of outcomes. Br Med Bull 97:47–80

40. Mayr HO, Reinhold M, Bernstein A, Suedkamp NP, Stoehr A (2015) Sports activity following total knee arthroplasty in patients older than 60 years. J Arthroplasty 30(1):46–49

41. Myklebust G, Maehlum S, Holm I, Bahr R (1998) A prospective cohort study of anterior cruciate ligament injuries in elite Norwegian team handball. Scand J Med Sci Sports 8:149–153

42. Narring F, Berthoud A, Cauderey M, Favre M, Michaud PA. Condition physique et pratiques sportives des jeunes dans le canton de Vaud. Institut universitaire de médecine sociale et préventive de Lausanne, 1998. www.iumsp.ch/Publications/pdf/RdS11.pdf. Accessed 17 Mar 2015

43. National Federation of State High School Associations (2002) 2002 High School Participation Survey. National Federation of State High School Associations, Indianapolis.http://www.nfhs.org/. Accessed 17 Mar 2015

44. Parkkari J, Pasanen K, Mattila KV, Kannus P, Rimpela A (2008) The risk for a cruciate ligament injury of the knee in adolescents and young adults: a population-based cohort study of 46 500 people with a nine year follow-up. Br J Sports Med 42(6):422–426

45. Pecina M, Bojanic I (2004) Overuse injuries of the musculoskeletal system, 2nd edn. CRC Press, Boca Raton/London/New York/Washington, DC

46. Prodromos CC, Han Y, Rogwowski J et al (2007) Ameta-analysis of the incidence of anterior cruciate ligament tears as a func- tion of gender, sport, and a knee injury reduction regimen. Arthroscopy 23(12):1320–1325.e6

47. Ristolainen L, Heinonen A, Waller B, Kujala UM, Kettunen JA (2009) Gender differences in sport injury risk and types of injuries : a retrospective twelve-month study on cross-country skiers, swimmers, long-distance runners and soccer players. J Sport Sci Med 8:443–451

48. Ruedl G, Webhofer M, Linortner I, Schranz A, Fink C, Patterson C, Nachbauer W, Burtscher M (2011) ACL injury mechanisms and related factors in male and female carving skiers: a retrospective study. Int J Sports Med 32(10):801–806

49. Sahlin Y (1990) Sport accidents in childhood. Br J Sports Med 24:40–44

50. Sallis RE, Jones K, Sunshine S, Smith G, Simon L (2001) Comparing sports injuries in men and women. Int J Sports Med 22:420–423

51. Schmidt B, Höllwarth ME (1989) Sportunfälle im Kindes- und Jugendalter. Z Kinderchir 44:357–362

52. Slauterbeck JR, Fuzie SF, Smith MP, Clark RJ, Xu K, Starch DW et al (2002) The menstrual cycle, sex hormones, and anterior cruciate ligament injury. J Athl Train 37:275–278

53. Sorensen L, Larsen SE, Röck ND (1996) The epidemiology of sports injuries in school aged children. Scand J Med Sci Sports 6:281–286

54. Taylor BL, Attia MW (2000) Sports-related injuries in children. Acad Emerg Med 7:1376–1382

55. Title IX of the Education Amendments of 1972 (Title 20 USC. Sections 1681–1688). Discrimination based on sex or blindness. Chapter 38, 1972. http://www.dol.gov/oasam/regs/statutes/titleix.htm. Accessed 17 Mar 2015

56. Tursz A, Crost M (1986) Sports-related injuries in children. A study of their charasteristics, frequency, and severity, with comparison to other types of accidental injuries. Am J Sports Med 14:294–299

57. Van Mechelen W (1997) The severity of sports injuries. Sports Med 24:176–180

58. Velin P, Four R, Matta T, Dupont D (1994) Évaluation des traumatismes sportifs de l'enfant et de l'adolescent. Arch Pediatr 1:202–207

59. Waldeen M, Haagglund M, Magnusson H, Ekstrand J (2011) Anterior cruciate ligament injury in elite football: a prospective three-cohort study. Knee Surg Sports Traumatol Arthrosc 19(1):11–19

60. Wojtys EM, Huston L, Boynton MD, Spindler KP, Lindenfeld TN (2002) The effect of menstrual cycle on anterior cruciate ligament in women as determined by hormone levels. Am J Sports Med 30:182–188

61. Yang J, Tibbetts AS, Covassin T, Cheng G, Nayar S, Heiden E (2012) Epidemiology of overuse and acute injuries among competitive collegiate athletes. J Athl Train 47:198–204

Causes of Overuse in Sports

3

Felix Fischer, Jacques Menetrey, Mirco Herbort,
Peter Gföller, Caroline Hepperger,
and Christian Fink

F. Fischer, MSc
Research Unit for Orthopedic Sports Medicine and
Injury Prevention, Institute of Sports Medicine,
Alpine Medicine and Health Tourism,
UMIT - The Health and Life Sciences University,
Hall in Tirol, Austria

FIFA Medical Centre of Excellence Innsbruck/Tirol,
UMIT - The Health and Life Sciences University,
Hall in Tirol, Austria
e-mail: felix.fischer@umit.at

J. Menetrey, MD
Swiss Olympic Medical Center, Hôpitaux
Universitaires de Genève, Faculté de médecine de
Genève, Genève, Switzerland

Centre de Médecine de l'appareil locomoteur et du
Sport – HUG, Hôpitaux Universitaires de Genève,
Faculté de médecine de Genève, Genève, Switzerland

Département de chirurgie, Hôpitaux Universitaires de
Genève, Faculté de médecine de Genève, Genève,
Switzerland
e-mail: jacques.menetrey@hcuge.ch

M. Herbort, MD
Department of Traumatology, Hand- and
Reconstructive Surgery, University Hospital
Muenster (UKM), Münster, Germany
e-mail: mirco.herbort@ukmuenster.de
http://www.traumacentrum.de

P. Gföller, MD
FIFA Medical Centre of Excellence Innsbruck/Tirol,
Innsbruck, Austria

Gelenkpunkt – Center for Sports and Joint Surgery,
Innsbruck, Austria
e-mail: p.gfoeller@gelenkpunkt.com

C. Hepperger • C. Fink, MD (✉)
Research Unit for Orthopedic Sports Medicine and
Injury Prevention, Institute of Sports Medicine,
Alpine Medicine and Health Tourism,
UMIT - The Health and Life Sciences University,
Innsbruck, Austria

FIFA Medical Centre of Excellence Innsbruck/Tirol,
Innsbruck, Austria

Gelenkpunkt - Center for Sports and Joint Surgery,
Innsbruck, Austria
e-mail: c.hepperger@gelenkpunkt.com;
c.fink@gelenkpunkt.com

Key Points

1. Reports for sports injuries caused by overuse vary between various studies, ranging from 30 % up to 78 %.
2. Chronic overuse injury is a multifactorial process with many other factors such as local hypoxia, diminished nutrition, impaired metabolism, and local inflammation.
3. Sports which include high-speed running, rapid movements, and full-body contact more typically show acute injury, whereas low-contact sports with long training sessions or the same movement repeated numerous times show more overuse injuries.
4. Rapid changes in the qualitative and quantitative aspects of training or extremely intensive training are training errors which can lead to overuse injury.
5. Sudden increases in training volume and changes in the type of training, especially running, are considered to be risk factors for overuse injury.

© ESSKA 2016
H.O. Mayr, S. Zaffagnini (eds.), *Prevention of Injuries and Overuse in Sports: Directory for Physicians,
Physiotherapists, Sport Scientists and Coaches*, DOI 10.1007/978-3-662-47706-9_3

3.1   Introduction

An increasing number of people are participating in sports, and recreational exercise has become part of most people's daily lifestyle [32]. Living an active lifestyle produces many benefits for health and well-being, e.g., reducing cardiovascular morbidity [27] and type 2 diabetes incidence [5] and preventing certain types of cancer [12]. However, as more people live an active lifestyle and participate in training and competition, more musculoskeletal system injuries occur [32]. Once an injury occurs, many physicians simply classify all symptoms as injury symptoms without differentiating between damage (overuse injury) and injury (acute injury) [32]. To distinguish between these two classifications, an injury occurs in a single acute action (acute injury), whereas damage appears after repeated action as the result of many repetitive minor insults (overuse injury) [32, 46]. Pecina and Bojanic summarize that "the main characteristic of an injury is acuteness, whereas damage has a chronic character" [32]. Overuse refers to any type of tissue injury (to muscle, tendon, ligament, etc.) resulting from ongoing repetitive movements or from static loading. Overuse injuries result when the destructive process of applied stress exceeds the reparative process [2]. In other words, if the tissue is not able to repair itself because of repetitive microtraumas, musculoskeletal overuse injuries occur. The following structures can be affected by overuse injury: tendons, bursae, cartilage, bone, and the musculotendinous unit [32]. Where a large number of tendons exist that play an important role in work and play, overuse injuries are quite common [32].

3.2   Genesis of Overload Damage

Reports for sports injuries caused by overuse vary between various studies, ranging from 30 % up to 78 % [11, 32] depending on the type of sport and study population. In general, there are two types of sports activities that lead to overload damage: "endurance activities and those associated with repetitive performances requiring skill, technique and power (such as hitting a tennis ball)" [15]. Sports including high-speed running, rapid movements, or full-body contact more typically result in acute injury, whereas overuse injuries are often found in aerobic low-contact sports that involve long training sessions [3, 46] such as running and swimming [16, 34, 36]. However, a large number of overuse injuries also occur in technical sports, in which the same movement is repeated numerous times [3]. During the 2008 Summer Olympic Games, 22 % of all reported injuries were overuse injuries [21]. In an epidemiologic study of 16 collegiate sports teams, more than one-quarter of all injuries were overuse injuries [46]. The incidence of overuse running injuries in recreational and competitive distance runners is somewhere between 27 and 70 % during a 1-year period, identifying running as one of the main causes of overuse injuries of the lower extremities during activity [11]. When marching long distances while carrying extra equipment loads without proper preparation, 78 % of US Marine Corps recruits were diagnosed with overuse injury [32]. Even in the nineteenth century, this form of injury was already known as "recruit's foot" [13]. Swimmers and throwers are more prone to overuse injuries in the upper extremities because a swimmer performs approximately 4,000 overhead strokes in one training session, totaling more than 800,000 overhead strokes during one season [32]. Taking into account that top class swimmers perform a high training volume to sustain an elite level, approximately 60 % suffer from overuse injury in the shoulder area [32, 41].

The stress-frequency relationship is shown in Fig. 3.1, where an increasing running or swimming distance would increase the number of repetitions made during training as more steps or strokes are performed. A greater distance covered means a higher frequency (provided that speed remained unchanged), placing the involved musculoskeletal structures further to the right on the stress-frequency curve. If the location on the curve moves further to the right, it requires lower stress impact for the structure to fall into the injury zone. With increasing distance, the possibility of entering the injury zone is more likely [18].

Repeating the same movement numerous times applies stress to the body; as a consequence, the involved tissue adapts by strengthening and thickening. If the applied stress becomes too great and the body system is not able to recover and adapt quickly enough, this overload to the system leads to microtraumatic injuries, which cause inflammation of the affected area, resulting in an overuse injury as a direct response.

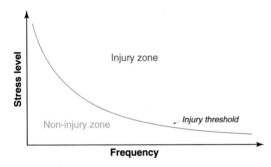

Fig. 3.1 Effect on overuse injury occurrence due to the theoretical relationship between stress application and frequency (According to Hreljac and Ferber [18])

Signs of overuse or inflammation include swelling, warmth, redness, or impaired function of the body part. Some or all of these signs may be present but not noticeable initially. Different classification systems that include four [4] to six stages [8] exist to define overuse injuries. In the four-stage system, the early stage of an overuse injury is discomfort that disappears during warm-up; the first sign is often stiffness or soreness. Continued use may cause further damage, leading to lasting pain through and beyond the warm-up and may worsen after finishing the activity. The third stage occurs when the discomfort gets worse during the activity. In the last stage, there is pain or discomfort all the time. Curwin and Stanish [8] distinguished six stages ranging from level 1, defined as having no pain and a normal level of sports performance, to level 6, which is characterized by pain during daily activities and an inability to perform in any sports. Increasing levels increase the description of pain and decreases the ability to perform any activity (Fig. 3.2).

There are several predisposing factors leading to overuse injuries of the musculoskeletal system, which consist of internal (intrinsic) and external (extrinsic) risk factors listed below (Fig. 3.3):

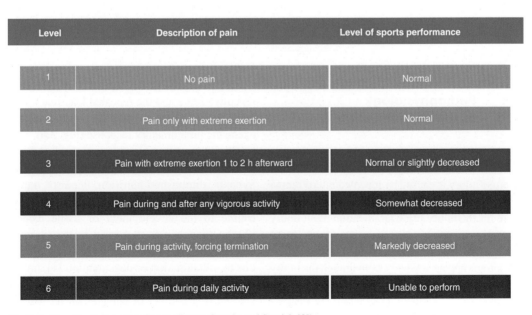

Level	Description of pain	Level of sports performance
1	No pain	Normal
2	Pain only with extreme exertion	Normal
3	Pain with extreme exertion 1 to 2 h afterward	Normal or slightly decreased
4	Pain during and after any vigorous activity	Somewhat decreased
5	Pain during activity, forcing termination	Markedly decreased
6	Pain during daily activity	Unable to perform

Fig. 3.2 Classification system (According to Curwin and Stanish [8])

Fig. 3.3 (a, b) Predisposing factors (According to Pecina and Bojanic [32])

a	**Internal (Intrinsic) risk factors**

Anatomical malalignment
- Leg length discrepancy
- Excessive femoral anteversion
- Knee alignment abnormalities (genu valgum, varum, or recurvatum)
- Position of the patella (patella infera or alta)
- Excessive Q-angle
- Excessive external tibial rotation
- Flat foot
- Cavus foot

Muscle-tendon imbalance of
- Flexibility
- Strenght

Other
- Growth
- Disturbances of menstrual cycles

b	**External (Extrinsic) risk factors**

Training errors
- Abrupt changes in intensity, duration, and/or frequency of training
- Poorly trained and unskilled athlete

Surface
- Hard
- Uneven

Footwear
- Inappropriate running shoes
- Worn-out shoes

3.3   Pathophysiology of Overload Damage

In contrast to traumatic injury, which is defined as injury resulting from a specific, identifiable event, overuse injury is caused by repeated microtrauma without a single, identifiable event that is responsible for the injury [3]. Classification of an injury as acute or overuse is obvious in most cases, but it is sometimes more difficult, particularly when the symptoms appear suddenly even though the injury is the result of a long-term process [3]. For instance, a stress fracture is caused by overuse over time [3]. Overload damage is the sum of repetitive force and the repeated load that affects the tissue. This process leads to microtraumas that excite the inflammatory response (Fig. 3.4).

Chronic overuse injury is a multifactorial process with many other factors such as local hypoxia, diminished nutrition, impaired metabolism, and local inflammation that may also contribute to the development of tissue damage [23]. Although it is discussed in the literature whether the pathogenesis of an overuse injury is due to inflammation or degeneration [1], injured tissue reacts with an inflammatory process, leading to a number of changes in the vascular net, blood, and connective tissue [32]. Pecina and Bojanic identify the inflammatory process as the primary cause of overuse injury, leading to degenerative changes during the course of its chronic stages [32]. If the inflammation is not fully cured, reinflammation may occur, which subsequently leads to chronic complaints. The pathophysiology of overload damage is a dynamic process, and overuse injuries are caused by repetitive, low-grade forces exceeding the tolerance of the tissues. The tissue can repair the damage without demonstrable clinical symptoms in most cases; however, in some cases, a permanent repeated workload meets or exceeds the ability of the tissue to repair and adapt, resulting in damage that leads to the symptoms of an overuse injury [45]. These repetitive microtraumas are the result of repeated exposure of the musculoskeletal tissue to low-magnitude forces [22], and this microtraumatic response is the failed adaption to physical load and use [32]. These effects are chronic and cumulate in microtraumatic injury [32]. For example, most elbow injuries result not from acute trauma but from repetitive microtrauma and chronic stress overload of the joint (Fig. 3.5) [31].

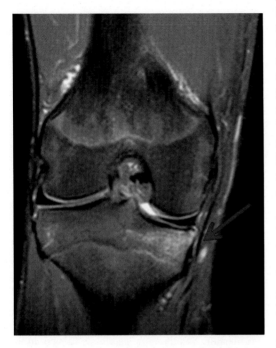

Fig. 3.4 Bone marrow edema of the medial tibial condyle due to overuse in a 17-year-old football player. *Red Arrow* indicate the injury/damaged tissue

3.4 Physical Constitution

The desired goal of improved performance is reached by prescribed periods of intense activity. During these periods of intense activity, stress is applied to the body. During periods of rest and recovery, the involved tissue adapts by strengthening and thickening. However, a mismatch between overload and recovery may lead to a breakdown on

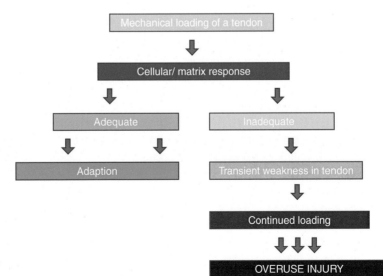

Fig. 3.5 Tendon response to mechanical loading [32]

Fig. 3.6 The four different zones of loading across a joint according to Dye et al. [9]

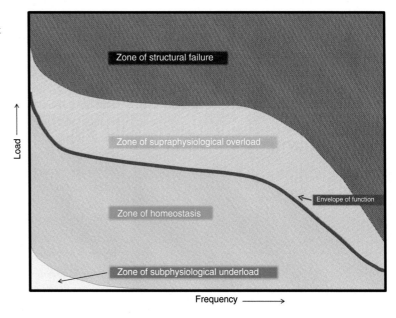

a cellular, extracellular, or systemic level [44]. Being a beginner to sports or starting again after a hiatus often accompanies poor preparation and/or inadequate conditioning that can substantially increase the incidence of overuse injuries [32]. This is often the case in young athletes, especially when they are beginners or in older athletes who want to restart training at an athletic level attained some time before the training interruption. Tissue breakdown and overuse injury is then the direct response at the cellular level when tissue that fails to adapt to new or increased demands is exposed to repetitive overload. Even before the person experiences pain and becomes symptomatic, subclinical tissue damage might have occurred. Rapid increases in training load and inadequate recovery may cause global overtraining syndrome at the systemic level [44].

Dye et al. classified four different zones of loading across a joint (see Fig. 3.6): the zone of subphysiological underload; the zone of homeostasis, which is limited through the envelope of function; the zone of supraphysiological overload; and the zone of structural failure [9]. Training load around the area of the envelope of function leads to adaptive processes that make tissues stronger and thicker, leading to enlargement of the zone of homeostasis; the envelope of function widens, and the athlete is able to perform under higher loads or achieve longer durations of physical activity. High load, high frequency, or combined overloading over a period of time pushes the body systems into the zone of supraphysiological overload or even structural failure, which subsequently leads to an injury. The zone of subphysiological underload affects an injured athlete who is not able to train or perform in any physical activity, causing catabolic processes, e.g., when extremities must be immobilized to support healing and therefore cannot be moved. In the same way that the envelope of function widens through training load, insufficient stimuli or injury decreases the envelope of function, and the zone of homeostasis is reduced. If an athlete returns to sports after an injury or a long period of inactivity, the envelope of function is lower than before and is likely significantly lower than recognized by the athlete. Returning to sports and restarting the prior training volume at which the athlete was accustomed to performing might exceed the envelope of function due to poorer physical constitution as a consequence of training interruption.

3.5 Gender Differences

There are differences between genders in the context of anatomic, physiologic, and psychological aspects [32]. Although athletic injuries seem to be sports rather than gender related [32], male athletes seem to have a higher risk of severe sports-related injuries [35], while women are more affected by overuse injury than men [6, 14, 37]. A study by Yang et al. revealed "that more than one-quarter of all injuries sustained by collegiate athletes were overuse injuries, with female athletes having a higher rate of overuse injuries than male athletes" [46]. Gender differences in the incidence of patellofemoral pain syndrome lead to a rate of 20 % in women versus 7.4 % in men, while the risk of repeated patellar dislocation is six times higher in women than in men [32]. Relative to body weight, women's limbs are shorter and smaller than those of men. Smaller bones are less powerful levers; additionally, women have a wider Q angle, which is a predisposing factor for patellofemoral problems [32]. Men have greater muscle mass than women [7, 20, 40], and women have decreased muscle-to-strength ratios between the quadriceps and hamstrings relative to men [32, 40]. As such, relative muscle weakness may be a risk factor for overuse injuries. Frequently conducted strength training to increase muscle power has been shown to decrease overuse injuries in the lower extremities of female recruits [33]. There are differences in muscle activity in response to anterior tibial translation, as female athletes use their quadriceps muscle, while male athletes and sex-matched nonathletic people rely more on their hamstring muscles for initial knee stabilization [19]. Increased ligamentous laxity is a widely accepted belief that is only partially true, while the role of cyclical estrogen and progesterone production has yet to be determined [32]. However, menstrual status among female athletes is considered to represent a risk factor for overuse injury [29, 36]. In young female athletes, intense training is often accompanied with low caloric intake, leading to amenorrhea or delayed menarche [28] that "may lead to hypoestrogenism and a decrease in bone material content predisposing these athletes to stress fractures" [32].

3.6 Mental Influences and Psychological Factors

Athletes who suffer overuse injury often mention psychological factors as a cause for their injury, e.g., "having too much of a drive," "wanting to succeed," "not knowing when to stop," and "training despite being fatigued" [45]. In certain sports such as high-performance bodybuilding, it seems that pain is accepted as part of the sports. Although the injury rate is low, a large number of athletes complain of pain or problems during routine workouts. High levels of motivation, discipline, and willpower may explain this phenomenon [39]. There is growing evidence that athletes' psychological characteristics are important for understanding overuse injuries in *sports* [42]. In a study by Timpka et al., 68 % of athletes reported an injury during the study period. Of these reported injuries, 96 % were classified as being associated with overuse [42]. Injury risk is increased by maladaptive coping through self-blame and hyperactivity [42]. Male adult athletes with previous serious injuries also showed an increased risk of injury, while the injury risk was decreased for male and female adult athletes without previous injury [42]. This finding may lead to the assumption that male athletes consider their body to be a tool that must function; training too hard and not granting the body enough rest during injury are more of a risk factor that can be attributed to male athletes. Attentiveness and care for the body during injury periods may decrease the risk of re-injury. The psychological factors of self-blame and hyperactivity, combined with previous injury, age, and gender, are considered to be risk factors for overuse injuries in athletes [42]. Not only should physical forces be considered but the importance of psychological factors and social context must be included as well because "an athlete with high physical demands may be more likely to suffer an overuse injury if she also feels herself to be in a stressful situation" [45]. If athletes return to sports after an injury, it has been shown that they experience injury-related distress although they are physically recovered from their injuries [43].

3.7 Excessive Work in Training and Competition

Sudden increases in training volume and changes in the type of training, especially running, are considered to be risk factors for overuse injury [32]. Achilles tendinopathy is the most common overuse injury in runners, followed by runner's knee, plantar fasciitis, and shin splints [25]. Knobloch et al. showed that running more than four times a week or an exposure for more than 2600 km led to a higher risk of shin splint overuse injuries. Runners with more than 10 years of experience had a higher risk of overuse injuries of the back and Achilles tendinopathy; training more than 65 km/week increased the risk for back overuse injuries [25]. Male runners over 35 years of age are more often prone to these complaints, especially after they have intensified their training routine with interval runs on the track or threshold runs (Fig. 3.7) [24].

In addition, the type of sports is related to injury. Football players had significantly more all-cause injuries (5.1 injuries/1000 exposure hours) than other sports (2.1–2.8/1000 exposure hours), while more runners than football players reported overuse injuries (59 % vs. 42 %) [34]. With increasing age, injury incidence increased during matches and decreased during training in elite youth football players [10], while the risk of suffering an overuse injury was two times higher during training than matches [10]. Combining data for 15 collegiate sports, injury rates were statistically significantly higher in games (13.8/1000 exposure hours) than in practices (4.0 injuries per 1000/ exposure hours) [16]. A study of elite football players showed an injury rate of 21.7/1000 match hours compared to 3.7/1000 training hours [26], while the 2001 FIVB (Federation Internationale de Volleyball) Beach Volleyball Injury Study showed that the incidence was 3.1 per 1000 competition hours and 0.8 per 1000 training hours [3]. In both sports, the injury rate was higher during matches than during training. Sports such as football, which include high-speed running, rapid movements, and full-body contact, more typically show acute injury, whereas low-contact sports with long training sessions or the same movement repeated numerous times [46], such as running and swimming [16, 34, 36], show more overuse injuries. High-speed running and rapid movements combined with full-body contact seem to be responsible for the higher rate of injury. A study of top-level endurance athletes showed that training-related risk factors for overuse injuries include a small number of recovery days and a large amount of training [36]. A previous history of injury, as well as walking or running more than 32.2 km per week are seen as strong predictors of overuse musculoskeletal injury [44]. Excessive mileage (>64 km/week) is a factor contributing to overuse injuries such as iliotibial band syndrome, bursitis, and stress fractures [30]. Excessive training with no recovery between training sessions leads to a 5.2-fold risk increase of an overuse injury in

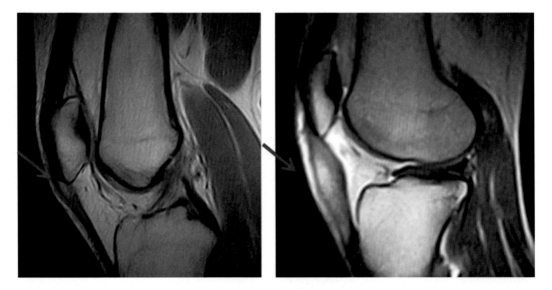

Fig. 3.7 Examples of patellar tendinopathy ("jumper's knee"). *Red arrows* are significant to indicate the injury/damaged tissue

athletes with less than 2 rest days per week during the training season. Athletes who trained more than 700 h during a year had a 2.1-fold risk increase for overuse injury compared to others [36]. Athletes' statements on excessive training, such as "not knowing when to stop" and "training despite being fatigued" or, in competition/ training, "having too much of a drive" and "wanting to succeed" [45], lead to the assumption that not only the physical forces but also the psychological factors and the social context could be of importance [45] for the coach and the athlete while planning and conducting training (Fig. 3.8).

Considering age as a risk factor, data shows that there is a greater number of tendon overuse injuries in older compared to younger athletes. This trend may indicate that age-related degeneration should be taken into account when assessing the cause of tendon injuries [36]. Older athletes should also be aware of the risk for an overuse injury when they are in the beginning of their training program or when they return to sports after a long training break.

3.8 Incorrect Training Methods

Training errors, such as not allowing enough resting time and fatigue, are among the most common causes of overuse injury [36]. Training errors cited in the literature include "running a distance that is too long, running at an intensity that is too high, increasing distance too greatly or intensity too rapidly, and performing too much uphill or downhill work" [44]. If the athlete has a history of prior injury, he/she should be checked for an inappropriate training approach, training errors, or inadequate technique [32]. Stress fractures of the bone are a typical overuse injury as a consequence of training errors in as many as 22–75 % of injury cases (Fig. 3.9) [32].

Rapid changes in the qualitative and quantitative aspects of training or extremely intensive training are training errors which can lead to overuse injury [32]. Another training error that leads to overuse injury, which is particularly common in running disciplines, is characterized by an excessively high mileage or "mileage

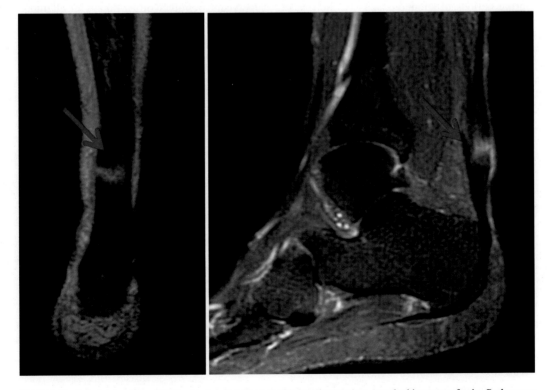

Fig. 3.8 Partial rupture of the Achilles tendon in a 45-year-old marathon runner – gradual increase of pain. *Red arrows* are significant to indicate the injury/damaged tissue

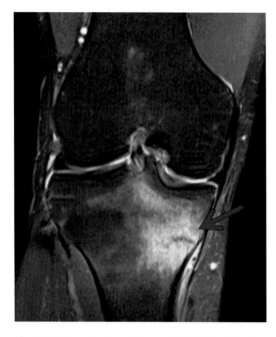

Fig. 3.9 Stress fracture of the medial tibial condyle in a 28-year-old female ultramarathon runner. *Red Arrow* indicate the injury/damaged tissue

mania" [30, 32, 36]. If rest periods are imposed during periods of strenuous activity, the risk of stress fracture decreases significantly [32, 36]. This reduction occurs because less load is applied to the bone, and the bone tissue has more time to repair microdamage [32, 38]. Hreljac reported that observations from clinical studies have estimated that over 60 % of running injuries could be attributed to training errors and suggests that all overuse running injuries are the result of training errors. This statement was made because if an individual has suffered an overuse running injury, he/she must have exceeded "her/his limit of running distance and/or intensity in such a way that the remodeling of the injured structure predominated over the repair process due to the stresses placed on the structure" [17]. Furthermore, the author states that each runner could have avoided these injuries by training differently based on individual limitations or, in some cases, by not training at all [17]. Understanding that there is a link between most overuse running injuries and training is important for the injured individual. With the correct advice, a modification of the training program can be made, especially when

the injurious aspects of the training regime are known [17]. As has been noted previously, sports with a large number of repetitive actions in training present greater risks of overuse injury; combined with poor technique in training sessions, the risk is even higher [44]. Through inadequate or poor technique, excessive pressure is applied to particular joints or muscles, which leads to overuse injury. For example, a poor backhand technique often causes tennis elbow, known as lateral epicondylitis [22]. Ignoring the first signs of an overuse injury caused by poor technique or incorrect training methods, e.g., levels 1–3 on the classification system according to Curwin and Stanish [8] (see Sect. 2.1), the injury is aggravated by maintaining the incorrect training method and could lead to a more severe state of injury. However, incorrect training methods are not only responsible for overuse injuries but can also lead to more acute injuries that are often common with the misuse of exercise machines or the use of excessive weight in barbell training. Variation of training routines, learning proper technique from the start, and training under the supervision of trained professionals are all important to avoid overuse and acute injuries.

References

1. Abate M, Silbernagel KG, Siljeholm C, Di Iorio A, De Amicis D, Salini V, Werner S, Paganelli R (2009) Pathogenesis of tendinopathies: inflammation or degeneration? Arthritis Res Ther 11:235
2. Archambault JM, Wiley JP, Bray RC (1995) Exercise loading of tendons and the development of overuse injuries. A review of current literature. Sports Med 20:77–89
3. Bahr R (2009) No injuries, but plenty of pain? On the methodology for recording overuse symptoms in sports. Br J Sports Med 43:966–972
4. Brenner JS (2007) Overuse injuries, overtraining, and burnout in child and adolescent athletes. Pediatrics 119:1242–1245
5. Bull FC, Armstrong TP, Dixon T, Ham S, Neiman A, Pratt M (2004) Physical inactivity. In: Ezzati M, Lopez AD, Rodgers A, Murray CJL (eds) Comparative quantification of health risks: global and regional burden of disease attributable to select major risk factors. World Health Organization, Geneva, p 729–881
6. Cuff S, Loud K, O'Riordan MA (2010) Overuse injuries in high school athletes. Clin Pediatr (Phila) 49:731–736

7. Cureton KJ, Collins MA, Hill DW, McElhannon FM (1988) Muscle hypertrophy in men and women. Med Sci Sports Exerc 20:338–344
8. Curwin S, Stanish W (1984) Tendinitis: its etiology and treatment. Collamore Press, Lexington
9. Dye S, Wojtys E, Fu F, Fithian D, Gillquist J (1998) Factors contributing to function of the knee joint after injury or reconstruction of the anterior cruciate ligament. J Bone Joint Surg Am 80:1380–1393
10. Ergün M, Denerel N, Binnet M, Ertat A (2013) Injuries in elite youth football players: a prospective three-year study. Acta Orthop Traumatol Turc 47:339–346
11. Ferber R, Hreljac A, Kendall KD (2009) Suspected mechanisms in the cause of overuse running injuries: a clinical review. Sports Health 1:242–246
12. Friedenreich CM, Neilson HK, Lynch BM (2010) State of the epidemiological evidence on physical activity and cancer prevention. Eur J Cancer 46:2593–2604
13. Geiger J, Rottenburger C, Uhl M (2010) Stressfrakturen bone stress injuries. Radiol up2date 10:35–50
14. Hart DA, Archambault JM, Kydd A, Reno C, Frank CB, Herzog W (1998) Gender and neurogenic variables in tendon biology and repetitive motion disorders. Clin Orthop Relat Res 351:44–56
15. Hess GP, Cappiello WL, Poole RM, Hunter SC (1989) Prevention and treatment of overuse tendon injuries. Sports Med 8:371–384
16. Hootman JM, Dick R, Agel J (2007) Epidemiology of collegiate injuries for 15 sports: summary and recommendations for injury prevention initiatives. J Athl Train 42:311–319
17. Hreljac A (2004) Impact and overuse injuries in runners. Med Sci Sports Exerc 36:845–849
18. Hreljac A, Ferber R (2006) A biomechanical perspective of predicting injury risk in running. Int Sport J 7:98–108
19. Huston LJ, Wojtys EM (1996) Neuromuscular performance characteristics in elite female athletes. Am J Sports Med 24:427–436
20. Janssen I, Heymsfield SB, Wang ZM, Ross R (2000) Skeletal muscle mass and distribution in 468 men and women aged 18–88 yr. J Appl Physiol 89:81–88
21. Junge A, Engebretsen L, Mountjoy ML, Alonso JM, Renström PAFH, Aubry MJ, Dvorak J (2009) Sports injuries during the Summer Olympic Games 2008. Am J Sports Med 37:2165–2172
22. Kannus P (1997) Etiology and pathophysiology of chronic tendon disorders in sports. Scand J Med Sci Sports 7:78–85
23. Kannus P, Paavola M, Paakkala T, Parkkari J, Järvinen T, Järvinen M (2002) Pathophysiology of overuse tendon injury. Radiologe 42:766–770
24. Knobloch K (2014) Laufsportverletzungen. Dominanz von Überlastungsschäden. Med Sports Netw 3:16–19
25. Knobloch K, Yoon U, Vogt PM (2008) Acute and overuse injuries correlated to hours of training in master running athletes. Foot Ankle Int 29:671–676
26. Kristenson K, Bjørneboe J, Waldén M, Andersen TE, Ekstrand J, Hägglund M (2013) The Nordic Football Injury Audit: higher injury rates for professional football clubs with third-generation artificial turf at their home venue. Br J Sports Med 47:775–781
27. Löllgen H, Löllgen D (2004) Physical activity and prevention of disease. Dtsch Med Wochenschr 129:1055–1056
28. Van de Loo DA, Johnson MD (1995) The young female athlete. Clin Sports Med 14:687–707
29. Nichols JF, Rauh MJ, Lawson MJ, Ji M, Barkai H-S (2006) Prevalence of the female athlete triad syndrome among high school athletes. Arch Pediatr Adolesc Med 160:137–142
30. Paluska S (2005) An overview of hip injuries in running. Sports Med 35:991–1014
31. Patten RM (1995) Overuse syndromes and injuries involving the elbow: MR imaging findings. Am J Roentgenol 164:1205–1211
32. Pecina M, Bojanic I (2004) Overuse injuries of the musculoskeletal system, 2nd edn. CRC Press, Boca Raton/London/New York/Washington, DC
33. Rauh MJ, Macera CA, Trone DW, Shaffer RA, Brodine SK (2006) Epidemiology of stress fracture and lower-extremity overuse injury in female recruits. Med Sci Sports Exerc 38:1571–1577
34. Ristolainen L, Heinonen A, Turunen H, Mannström H, Waller B, Kettunen JA, Kujala UM (2010) Type of sport is related to injury profile: a study on cross country skiers, swimmers, long-distance runners and soccer players. A retrospective 12-month study. Scand J Med Sci Sports 20: 384–393
35. Ristolainen L, Heinonen A, Waller B, Kujala UM, Kettunen JA (2009) Gender differences in sport injury risk and types of injuries : a retrospective twelve-month study on cross-country skiers, swimmers, long-distance runners and soccer players. J Sports Sci Med 8:443–451
36. Ristolainen L, Kettunen JAJ, Waller B, Heinonen A, Kujala UM (2014) Training-related risk factors in the etiology of overuse injuries in endurance sports. J Sports Med Phys Fitness 54:78–87
37. Sallis RE, Jones K, Sunshine S, Smith G, Simon L (2001) Comparing sports injuries in men and women. Int J Sports Med 22:420–423
38. Scully T, Besterman G (1982) Stress fracture: a preventable training injury. Mil Med 147:285–287
39. Siewe J, Marx G, Knöll P, Eysel P, Zarghooni K, Graf M, Herren C, Sobottke R, Michael J (2014) Injuries and Overuse Syndromes in Competitive and Elite Bodybuilding. Int J Sports Med 35:943–948
40. Smith FW, Smith PA (2002) Musculoskeletal differences between males and females. Sports Med Arthrosc Rev 10:98–100
41. Taunton JE, McKenzie DC, Clement DB (1988) The role of biomechanics in the epidemiology of injuries. Sports Med 6:107–120
42. Timpka T, Jacobsson J, Dahlstrom O, Kowalski J, Bargoria V, Ekberg J, Nilsson S, Renstrom P (2014)

Psychological risk factors for overuse injuries in elite athletics: a cohort study in Swedish youth and adult athletes. Br J Sports Med 48:666

43. Verhagen EALM, van Stralen MM, van Mechelen W (2010) Behaviour, the key factor for sports injury prevention. Sports Med 40:899–906

44. Wilder RP, Sethi S (2004) Overuse injuries: tendinopathies, stress fractures, compartment syndrome, and shin splints. Clin Sports Med 23:55–81

45. Van Wilgen CP, Verhagen EALM (2012) A qualitative study on overuse injuries: the beliefs of athletes and coaches. J Sci Med Sport 15:116–121

46. Yang J, Tibbetts AS, Covassin T, Cheng G, Nayar S, Heiden E (2012) Epidemiology of overuse and acute injuries among competitive collegiate athletes. J Athl Train 47:198–204

General Prevention Principles of Injuries

4

Stefano Zaffagnini, Federico Raggi,
Jorge Silvério, Joao Espregueira-Mendes,
Tommaso Roberti di Sarsina, and Alberto Grassi

Key Points

1. An effective injury prevention can not only include the proper equipment and training, but should also cover lifestyle and the mental condition of the athlete.
2. Knowledge of the correct training methods and respect of the recovery time by athletes and coaches play a crucial role in prevention.
3. Athletes must follow an appropriate diet to fulfill their functional requirements and adopt a healthy lifestyle, avoiding excesses and voluptuary habits.
4. Body and mind can not be considered separately: proper psychological preparation helps the athlete to bear the agonistic stress and physical loads.
5. The proper equipment and the appropriate structures are the final major component of the process of injury prevention.

4.1 Introduction

General prevention of injuries is a fundamental part to be completed and respected for all people involved in sports activity.

The athletes, the coaches, the physical trainer and the team manager should all be aware of these important steps that allow a safe sports activity reducing the risks of temporary invalidating injuries.

General prevention includes several aspects that can be considered intrinsic and extrinsic. The intrinsic ones are related to training methods, nutrition, way of living and psychological attitudes. The extrinsic ones include active and passive protection of the athletes and the features of the playing field in order to reduce the risk of injuries.

In this chapter, all those aspects will be evaluated and discussed.

4.2 Training Methods

Improving performance is not just about training; athletes need to follow a carefully planned training programme.

This programme must be systematic and take into account the demands of activity and the needs, preferences and abilities of the athlete.

There are a number of principles that athletes and coaches must follow to improve performances and to work in safety.

S. Zaffagnini (✉) • F. Raggi, MD
T. Roberti di Sarsina, MD • A. Grassi, MD
Clinica Ortopedica e Traumatologica II, Laboratorio di Biomeccanica ed Innovazione Tecnologica, Istituto Ortopedico Rizzoli,
Via di Barbiano, 1/10, 40136 Bologna, Italy
e-mail: stefano.zaffagnini@biomec.ior.it

J. Silvério • J. Espregueira-Mendes
Clínica do Dragão – Espregueira Mendes Sports Centre – FIFA Medical Centre of Excellence,
Porto, Portugal

© ESSKA 2016
H.O. Mayr, S. Zaffagnini (eds.), *Prevention of Injuries and Overuse in Sports: Directory for Physicians, Physiotherapists, Sport Scientists and Coaches*, DOI 10.1007/978-3-662-47706-9_4

Overload: In order to progress and improve performances, the human body must be subjected an additional stress. Doing this will cause long-term adaptations, enabling athlete's body to work more efficiently to cope with this higher level of performance. Overloading can be achieved by following the acronym FITT:

- Frequency: Increasing the number of training session weekly
- Intensity: Increasing the load of exercise. For example, running at 12 km/h instead of 10 or increasing the weight you are squatting with
- Time: Increasing the length of time for each session. For example, cycling for 45 min instead of 30
- Type: Increase the difficulty of training. For example, progress from walking to running

Specificity: The type of training should be sport specific. The energy system used predominantly (i.e. strength, aerobic endurance) and the fitness and skill components most important to the specific sport should be trained: for example, agility, balance or muscular endurance.

Athletes and coaches should also test the components important in their specific sport to find strengths and weaknesses.

Reversibility: All improvements, made by an athlete, will be reversed if he stops training. After an injury stop, or the seasonal break, the athlete may not be able to resume training at the point he left off. In some sports, a period even as little as a week can be critical. In those cases, the training programme must be modified.

Variance: Variations in training are fundamental to keep the athletes interested and to give different urges to the body. The training should not be repetitive as not to overload the same groups of muscles and not to be psychologically stressful.

Usually, professional athletes will play a completely different sport in between their main season, to keep their fitness up while still having a rest.

4.2.1 Periodization

Training should be organised and planned in advance of a competition or performance. It should consider the athlete's potential, his performance in tests or competition and the calendar.

The basic building block of training periodization, usually the training week, is called the microcycle. Microcycles form the building blocks for a discrete unit of training, usually a few weeks in duration, termed a mesocycle. A number of repeated mesocycles form a macrocycle.

The macrocycle: A macrocycle refers to an annual plan that works towards peaking for the goal competition of the year. There are three phases in the macrocycle: preparation, competitive and transition.

The entire preparation phase should be around 2/3–3/4 of the macrocycle. The preparation phase is further broken up into general (to build the aerobic base) and specific (to be more efficient on the final format of the sport) preparation of which general preparation takes over half.

The competitive phase consists in several competitions that lead up to the main competition and are very useful in order to test the athlete's conditions with some specific tests. Testing might include performance level, different types of equipment, prerace meals, ways to reduce anxiety before a race or the length needed for the taper phase. This phase ends with the main competition of the season.

The transition phase is important for physical and psychological reasons; it consists in a break after a year of training.

The mesocycle: A mesocycle represents a phase of training with a duration of between 2 and 6 weeks or microcycles, but this can depend on the sporting discipline. A mesocycle can also be defined as a number of continuous weeks where the training programme emphasises the same type of physical adaptations, for example, muscle mass and anaerobic capacity. During the preparatory phase, a mesocycle commonly consists of 4–6 microcycles, while during the competitive phase, it will usually consist of 2–4 microcycles depending on the competition's calendar.

The goal of the plan is to fit the mesocycles into the overall plan timeline-wise to make each mesocycle end on one of the phases and then to determine the workload and type of work of each cycle based on where in the overall plan the given mesocycle falls. The goal in mind is to make sure that the body peaks for the high priority competitions by improving each cycle along the way.

The microcycle: A microcycle lasts typically a week because of the difficulty in developing a training plan that does not align itself with the weekly calendar. Each microcycle is planned based on where it is in the overall macrocycle.

A microcycle is also defined as a number of training sessions, built around a given combination of acute programme variables, which include progression as well as alternating effort (heavy vs. light days). The length of the microcycle should correspond to the number of workouts – empirically often 4–16 workouts – it takes for the athlete to adapt to the training programme. When the athlete has adapted to the programme and no longer makes progress, a change to one or more programme variables should be made [51].

Each workout must be preceded by a suitable warm-up and must end with a cooling down phase, taking particular care to the elasticity of muscles.

A good warm-up prepares the body for a more intense activity. It increases heartbeat, raises muscle temperature and increases breathing rate. Warming up gives to the athlete time to adjust to exercise demands. This can improve performance and reduce injury risk.

Stretching helps the body get ready for exercise and should be performed before and after each exercise session. It increases flexibility, improves to perform movements, reduces risk of injury and also gives mental relaxation benefits to the athlete.

As the warm-up prepares the body for the workout, an effective cooldown gives to the body time to recover.

4.2.2 Warm-Up

The term warm-up in sport is defined as a period of preparatory exercise in order to enhance subsequent competition or training performance [27].

The ideal warm-up will depend on the sport, the level of competition and the age of the participants. The warm-up should incorporate the muscle groups and activities that are required during training or competition. The intensity of the warm-up should begin at a low level gradually building to the level of intensity required during training or competition. For most athletes, 5–10 min is enough. However, in cold weather, the duration of the warm-up should be increased.

An appropriate warm-up consists of the three different factors, as recommended by Safran et al. [53].

These three factors are the components of a warm-up that are most commonly cited in the warm-up literature and include:

1. A period of aerobic exercise to increase body temperature
2. A period of sport-specific stretching to stretch the muscles to be used in the subsequent performance
3. A period of activity incorporating movements similar to those to be used in the subsequent performance.

Fradkin, Gabbe and Cameron reviewed five high-quality studies that investigated the effects of warming up in humans on injury risk in physical activity. Three of the studies found that performing a warm-up prior to physical activity significantly reduced the risk of injury, while the remaining two studies found that the warm-up was not effective in reducing the number of injuries [72].

The studies that found that warming up prior to physical activity reduced the risk of injury investigated handball and American football. Those studies found that by warming up, the number of traumatic and overuse injuries in the intervention group was significantly lower than in the control group.

Currently, there are many prevention programmes for specific sport or anatomical areas.

An example of preventive warm-up programme widely used in football is the FIFA 11+, which was developed in cooperation with national and international experts under the leadership of the FIFA Medical and Research Centre (F-MARC), to reduce the incidence of football injuries.

The FIFA 11+ is a simple and easy warm-up programme to implement sports injury prevention and is comprised of ten structured exercises. The programme includes exercises that focus on core stabilisation, eccentric training of thigh muscles, proprioceptive training, dynamic stabilisation and plyometric drills performed with good postural alignment. The evidence for including a plyometric component, as a portion of a ligament injury prevention programme, is relatively strong. A systematic review of the literature found that reduced ACL injury risk occurred in those interventions that included plyometrics as part of the training programme, while in those that did not include plyometrics it did not reduce ACL injury risk [28].

The FIFA 11+ programme requires no technical equipment other than a ball and after familiarisation can be completed in 10–15 min.

A recent review demonstrated how the FIFA 11+ programme can decrease the incidence of injuries in both male and female amateur football players and also improve neuromuscular performance, enough to consider this programme a fundamental public health intervention [5].

Fig. 4.1 The figure of four, an example of static stretching

4.2.3 Stretching

Stretching is widely used by many athletes before exercise training and competition [55] because it increases range of motion (ROM). It is commonly believed that increasing ROM contributes not only to injury prevention but also to improved athletic performance [1].

Stretching techniques vary according to the type of sport, programme type and personal preference.

The three most common stretching methods are static, ballistic and proprioceptive neuromuscular facilitation (PNF) stretching.

Static stretching is a passive movement to the muscles' maximum ROM and maintaining that position for an extended time period. Static stretching is the type of stretching that most individuals perform and are familiar with (Fig. 4.1).

Static stretching increases range of motion (ROM) and can also decrease musculotendinous stiffness, even during short-duration (5–30 s) stretches [73].

A systematic review by Kay and Blazevich has showed that longer stretch durations (60 s)

are more likely to cause a small or moderate reduction in performance; however, a static muscle stretching totalizing 45 s can be used in pre-exercise routines without risk of significant decreases in strength-, power- or speed-dependent task performances [34].

Ballistic stretching consists of repetitive bouncing movements at the limit of range of motion using the stretched muscles as a spring which pulls out from the stretched position. This particular stretch increases ROM but also is associated with reduced muscle strength and can cause injury. This should be done with extreme caution and after pondering many factors (age, physical condition, experience) of the considered individual.

Proprioceptive neuromuscular facilitation stretching is a reflex activation and inhibition of agonist and antagonist muscles, resulting in increased range of motion. This stretching method can furthermore be subdivided into passive and active techniques. In the passive techniques, the target muscle is placed into a position of stretch followed by a static contraction. The muscle is then passively moved into a greater position of stretch [13]. In the active technique, the final passive stretch is exchanged by an active contraction of the antagonist, which stretches the target muscle [13]. Both PNF methods produce an ROM improvement decreasing stiffness of the tendon structures [36].

Fig. 4.2 (**a**, **b**) A CrossFit class during the cooling down phase

4.2.4 Cooling Down

Cooldown phase consists in gentle exercises after vigorous physical activity. It should include 5–10 min of slow cardio activity (jogging or walking) followed by 5–10 min of static stretching exercises.

This phase is recommended because it has been observed that cooling down helps in the removal of lactic acid, decreasing muscle soreness [16].

Cooling down also brings quickly the body from an exercise state back to a state of rest. It reduces heart and breath rates, gradually cools body temperature and restores to baseline the physiological systems (Fig. 4.2).

4.3   Way of Living

4.3.1   Nutrition

Not only an adequate diet improves athletes' performance but every single athlete's aspect of life can benefit from a good nutrition programme.

An athlete's energy need and nutrient requirement depend on various individual factors as weight, height, age, sex and metabolic rate and on the type, intensity, frequency and duration of training.

Every athlete or coach must calculate the adequate calorie and essential nutrient intake, according to the type of training, to reach good performances and to maintain good health.

Ideally, athletes need to take 60–70 % of their total calories from carbohydrate, and the remaining calories should be obtained from proteins (10–15 %) and fats (20–30 %).

Like the general population, it's a good practice that calories and nutrients come from a wide variety of foods on a daily basis, as reported in the food guide pyramid [21].

4.3.1.1 Carbohydrates as Fuel

Carbohydrates are the precursor needed to form muscle glycogen stores necessary for the physical exercise. If the glycogen stores level drops down, like during an endurance exercise, high intensity cannot be maintained and the athlete will feel exhausted.

For this reason, the right consumption of carbohydrates is important during the pre-workout and post-workout and also during exercise in some cases.

Before exercise, a proper meal is necessary to maintain optimal level of blood glucose for muscles. Carbohydrate intake before exercise can help to restore suboptimal liver glycogen stores, which the athlete will use during prolonged training and competition.

The pre-workout meal should be high in carbohydrates, not rich in fat, and readily digestible. Current researches suggest that 1–4 g of carbohydrate per kilogramme of body weight should be consumed 1–4 h before exercise [56]. High-fat or high-protein food should be avoided for their long gastric emptying time.

During an endurance exercise (longer than 90 min), muscles run out of glycogen, and they begin to take up glucose from the blood stream. Supplies of blood glucose can be drawn from liver glycogen, but with the persistence of the exercise, liver glycogen stores can be depleted and blood glucose level drops down. In these conditions, the athlete will note local muscular fatigue and will have to reduce his exercise intensity and in some cases it is possible to observe typical hypoglycaemia symptoms. The available evidence suggests that athletes should take in 25–30 g of carbohydrate every 30 min during an endurance exercise to maintain an adequate blood glucose level [14]. This amount can be obtained through either carbohydrate-rich foods or fluids during training (sports drink or energy bars).

After exercise, muscle glycogen that has been used must be resynthesised. It has been demonstrated that when carbohydrates (about 100 g) are consumed immediately after exercise, glycogen resynthesis can be faster [32]. It is also possible to obtain a more efficient glycogen resynthesis rate by adding protein to the post-workout meal. The reason for this improvement is that proteins can produce a greater insulin response and therefore activate glycogen synthase [71]. The current recommendations are to consume around 100 g of carbohydrates within a 30-min window postexercise to maximise muscle glycogen synthesis.

4.3.1.2 Proteins

The protein requirements do vary with the type and intensity of exercise performed and the total amount energy consumed. Several studies suggest that individuals who exercise at a higher intensity have a greater protein need [42, 61].

It has also been shown that the protein need depends on the energy intake; in fact, if calories from carbohydrates are insufficient, proteins will be used as an energy source, therefore increasing the need of protein intake [9].

Even if high-level athletes require a higher protein intake, being western diet often hyperproteic, general population doesn't need a protein supplement. This means that if the caloric intake is adequate, athletes don't need more proteins than the rest of the population.

Consuming more proteins than the amount that body can use should be avoided. When athletes follow diets that are high in protein, they can compromise their carbohydrate status and therefore affect their ability to train and compete at peak performance.

The unjustified large use of protein or amino acid supplements, common in some sports, can lead to dehydration, loss of urinary calcium, weight gain, and stress to the kidney and liver [59].

4.3.1.3 Fat

Fat also provides energy for exercise. Fat is the most concentrated source of food energy and supplies more than twice as many calories (9 kcal/gm) by weight as protein (4 kcal/g) or carbohydrate (4 kcal/g). Fat is the major, if not most important, fuel for light- to moderate-intensity exercise. Although fat is a valuable metabolic fuel for muscle activity during longer-term aerobic exercise, no attempt should be made to consume more fat.

However, it has been demonstrated that athletes who consumed a lower-fat, higher-carbohydrate diet had significantly more power and speed because of their higher muscular glycogen level [58].

4.3.1.4 Drugs

Athletes can easily fall under pressure because their performance is constantly being judged and evaluated. In addition, today at all levels, a certain number of athletes wants to win at all costs. Consequently, there is a big interest among some athletes in investigating drugs that may enhance performance. Those athletes believe that certain drugs may increase athletic performance and consider their use despite the well-known long-term health hazards.

One of the most common substances being abused by athletes is anabolic steroid. These chemical derivatives of testosterone are used by athletes to increase muscle mass, which improves strength and power. What sportsmen overlook are the long-term side effects of anabolic steroids including liver cancer, prostate cancer, abnormal sperm production and some psychological changes like aggressive and violent behaviour [70].

Another traditional doping agent is erythropoietin (EPO) that can increase blood cell count.

Having more red blood cells to carry oxygen, athletes aerobic capacity is improved and fatigue delayed. EPO injections increase haematocrit level, which may cause a heart overload. This is particularly dangerous when the heart rate slows down, such as during sleep. The increased thickness, or viscosity, of the blood increases the risk of blood clots, heart attacks and strokes.

At last, weight loss drugs are frequently taken by athletes to lose weight quickly to make a particular weight classification. Currently, the most common drug being used to lose weight fast is the ephedra. This product is commercialised as a dietary supplement for the purpose of weight loss. These compounds increase heart rate, constrict blood vessels (increasing blood pressure) and expand bronchial tubes (making breathing easier). Their thermogenic properties cause an increase in metabolism, thus leading to an increase in body temperature.

The use of this substances can cause potentially fatal side effects, including stroke, heart attack, seizures, or severe mental disorders.

4.3.2   Sleep Hygiene

Although the function of sleep is not fully understood, it is generally accepted that it is necessary to recover from previous wakefulness and prepare for functioning in the subsequent wake period.

Sleep deprivation can have significant effects on athletic performance, especially submaximal, prolonged exercise [49].

Compromised sleep may also influence learning, memory, cognition, pain perception, immunity and inflammation [11]. Furthermore, changes in glucose metabolism and neuroendocrine function as a result of chronic, partial sleep deprivation may result in alterations in carbohydrate metabolism, appetite, food intake and protein synthesis [60].

Data from some studies which investigated about the effect of sleep on performance suggest that increasing the amount of sleep in athletes may significantly enhance performance.

Mah et al. [39] instructed six basketball players to obtain as much extra sleep as possible following 2 weeks of normal sleep habits. Faster sprint times and increased free-throw accuracy were observed at the end of the sleep extension period. Mood was also significantly improved, with increased vigour and decreased fatigue. The same research group also increased the sleep time of a group of swimmers to 10 h per night for 6–7 weeks and reported that 15-m sprint, reaction time, turn time and mood were improved [39].

Moreover, athletes suffering from some degree of sleep loss may benefit from a brief nap, particularly if a training session is to be completed in the afternoon or evening. Following a 30-min nap, 20-m sprint performance was increased (compared to no nap), alertness was increased and sleepiness was decreased. In terms of cognitive performance, sleep supplementation in the form of napping has been shown to have a positive influence on cognitive tasks. Naps can markedly reduce sleepiness and can be beneficial when learning skills, strategy or tactics if compared to sleep-deprived individuals [48]. Napping may be beneficial for athletes who have to routinely wake early for training or for competition and for athletes who are experiencing sleep deprivation [67].

Athletes should focus on utilising good sleep hygiene to maximise sleep time. Strategies for good sleep include:

The bedroom should be cool, dark and quiet. Eye masks and earplugs can be useful, especially during travels.

Create a good sleep routine by going to bed at the same time and waking up at the same time.

Avoid watching television in bed, using the computer in bed and avoid watching the clock.

Avoid caffeine approximately 4–5 h prior to sleep (this may vary among individuals).

Do not go to bed after consuming too much fluid as it may result in waking up to use the bathroom.

Napping can be useful; however, generally, naps should be kept to less than 1 h and not too close to bedtime as it may interfere with sleep [24].

4.4 Mental Constitution

The practice of sport, particularly at higher levels, is not devoid of perils. One of these perils is the lesion. Most of the lesions suffered by athletes are caused by the high demands of the competitive sport: attaining great results in a short time period, lack of psychological preparation, growing physical loads, disrespect of the adaptation time that the body needs to adapt to progressively higher demands, etc. [17].

We cannot say that there is a particular type of personality more prone to be injured [57, 68] because the research so far has failed to prove that particular personality characteristics are associated to injuries. However, a sensation-seeking person can be more willing to practise higher-risk sports such as skiing, motorbike, escalade, etc. The probability of these sports that cause lesions is higher than others [17], and his practitioners having consciousness of this must take more precautions without putting in question his achievements. According to Malina et al. [40], we can divide the risk factors for lesions in internals and externals. In the first category, we can put physical factors, problems of structural alignment, lack of flexibility, lack of muscular strength, a poor capacity development, behaviour factors, lesion history and variations associated with the maturity. In the external risk factors, we must consider inadequate rehabilitation from a previous injury, training errors, game conditions, equipment, mixture of age groups, trainer's behaviours, parenting behaviours and the sports organisation.

What we say to the athletes with whom we work is that a physical injury always implies a psychological one also! So, if we treat a lesion from the physical side, we must treat it also from the psychological point of view. And with the advantage of some psychological techniques such as mental imagery, it can accelerate the recuperation time [68].

And, as the advances in sports medicine allow to an extraordinary reduction of the recuperation times, more and more important is the role of the psychological recuperation especially if we talk about professional sport where the athlete is a "financial investment" and must be fully recovered.

Initially, the researchers agreed on a five-stage process (similar to the one occurring in grief): after an injury, the athlete passes by different stages: immediately after the lesion the negation phase: "this has not happened to me". After he gains consciousness, the rage phase succeeds: "why me?" Then he starts seeking reasons for suffering the injury, and if he surpasses all these phases with success, the fourth stage is depression. Finally, the last phase is acceptance. The time an athlete spends in each of these phases and the velocity which he passes through each depends on his personality and the support he receives from the others [25]. Now the common understanding is that the injury experience is not so rigid and ordered. So, we can talk about three general categories of responses [65]:

1. Injury-relevant information processing: the injured athlete gathered information related to the injury and recognises the negative consequences.
2. Emotional upheaval and reactive behaviour: as soon as the athlete realises that he is injured, he experiences a set of emotions such as shock, disbelief, denial, isolation and self-pity.
3. Positive outlook and coping: it's when the athlete accepts the injury and deals with it starting coping efforts, has a good attitude and is relieved to see progression in his recuperation.

Athletes can also display other reactions [47]:

1. Identity loss: his self-concept as athlete becomes diffusing because of the effects of the injury.
2. Fear and anxiety: the thoughts about reinjuring and the worries about their recuperation and if some teammate replaces them are overwhelming. And all these can increase because normally an injured athlete has plenty of time.
3. Lack of confidence: as an injured athlete can't practise and play, they can lose confidence in their abilities.
4. Performance decrements: because of the diminished confidence and the lost practice time, athletes can experience performance decrements when they return to practice and competitions.

The consequences of an injury are not only in terms of pain and suffering but also in self-esteem, quality of life, stress and anxiety, anger, treatment compliance issues, depression, concentration and attention problems, mood alterations, sleep disturbance, and alterations in cognitive processes and directly or indirectly, through all these effects, can affect relationships.

Fortunately, today, the traditional thinking of considering a lesion a mere biological problem has been surpassed, and the role of the psychological factors and their treatment is widely recognised.

We can divide the psychological intervention on injuries in two stages: prevention and rehabilitation.

At the preventive level, psychologists can intervene in athletes at injury risk: the ones with negative life stresses, an increase in daily hassles, previous injuries and poor coping resources [8]. Moreover, if an athlete experiences anger and aggression during a competition or has competitive trait anxiety, the risk of injury increases. "The combination of stress history, poor coping resources and personality factors results in what theorists call an elevated stress response. This involves increased muscle tension, increased distractibility, and a narrowing of attention so that the athlete is not as aware of or responsible to critical events or cues. Prolonged exposure to stress also changes the body's endocrine system, making a person more susceptible to illness and slowing down the healing process" ([8], pp. 218).

Palmi [46] talks about three risk factors, the medical–physiological, psychological and sports related, and consequently suggests four intervention strategies:

1. Improving the basic information given to the athlete about the risk factors, the best preparation and the habits to prevent injuries.
2. Learning psychological resources to reduce the probability of injuries such as relaxation after finishing hard training sessions.

3. Planning the training and the competition with realistic goals adapting to his own condition and in order to avoid overtraining.
4. Improving technical resources because how much an athlete is ready to do a task is less the probability to get injured.

On the rehabilitation side, a psychologist intervention can address these aspects:

– Establish a good relationship of confidence with the athlete.
– Educate the athlete about the injury and the recovery process.
– The athlete must accept his responsibilities in the treatment trying to actively seek information about the lesion and what he can do in order to recover as soon as possible.
– The athlete must define clear, specific and realistic goals with a time frame in a short, medium and long term for the rehabilitation process and for the return to practices and competition.
– The athlete should know how to get support from friends, family, teammates and trainers.
– The athlete must work mentally in his recovery process taking consciousness of his attitude towards the injury and building positive self-talk about the recovery process, and if he can't practise physically, he can mentally rehearse and practise.
– The athlete can't isolate himself from friends and family.
– Teach the athlete how to cope with the possible setbacks.

The other sports medicine personnel involved in the rehabilitation process can act in order to [68]:

– Educate and inform the athlete about the injury and recuperation process
– Use appropriate motivation
– Demonstrate empathy and support
– Have a supportive personality (warm, open and not exaggerate confident)
– Foster positive interaction and customise training

– Demonstrate competence and confidence
– Encourage the athlete's confidence

4.5 Expectations of the Athletes and the Environment

One of the toughest things for an injured athlete to manage is with the self-expectations and also with the expectations from the others (family, trainers, managers, teammates, supporters and friends) about his recovery and returning to the practices and competitions.

One of the most important words we say to the injured athletes we work with is patience. Patience is very important for the athlete not to return too early and get reinjured but also to all the people around him who is important to the recovery process.

One athlete who is capable of working all the points mentioned earlier in the previous topic and who has a team of sports medicine personnel involved and is knowledgeable and has patience is far more near to recover well and return to his previous level of achievement or better!

As coach John Wooden said "Success is peace of mind which is a direct result of self-satisfaction in knowing you did your best to become the best you are capable of becoming"!

4.6 Active and Passive Protection

4.6.1 Active Protection

According to the definition of active approach in preventive medicine by Haddon [74], active protection from injury requires individual intervention to function correctly. These efforts involve educating the public on the safest courses of action and involve changing people's behaviour to increase their safety.

Active protection in sports injury prevention is represented by the proper equipment, which the player must wear, that may be helpful in many contact sports or sports where there is a risk of impact with another player or the ground.

4.6.1.1 Helmets

Helmets can reduce the two primary head injury risks which are concussion and severe head injury.

In some sports, such as US football, soccer, rugby and Australian football, concussion is the main brain injury risk. In other sports, there are severe head injury risks in addition to concussion, for example, equestrian and pedal cycling. In projectile sports, such as cricket, baseball, lacrosse and ice hockey, orofacial injury is a significant concern and has led to protective visors being worn attached to helmets.

In alpine sports, bicycling helmets have been shown to be effective in reducing moderate to severe head injuries. In equestrian sports, it is challenging to measure helmet effectiveness or efficacy, because all amateur and professional jockeys wear a helmet. In Australian football and rugby, there is no evidence that padded headgear prevents head and brain injury [41].

In a comprehensive review, Biasca et al. reported that the combination of increased helmet wearing and rigorous helmet standards had reduced the incidence of fatal and serious head injury in ice hockey, although concussion was a growing concern [7].

Types of Helmet Technologies

There are many types of sports helmets on the market; however, the two most important categories are single-impact helmets (or crash helmets) and multiple-impact helmets [29].

Single-impact helmets these helmets are typically designed to absorb a high-energy single impact in sports like motorsports, bike and alpine ski. The single-impact helmet is typically constructed of an outer shell designed to help distribute the force of the impact and an inner energy-absorbing liner. The energy-absorbing liner is often made of a material such as expanded polystyrene that deforms plastically under impact and thus releases energy. This means that after one impact, the helmet is compromised and lacks much of its original ability to manage impact energies. In terms of concussion, expanded poly-styrene liners do very little to manage accelerations below concussion thresholds and are particularly poor at reducing rotations associated with concussions [30].

Multiple-impact helmets these helmets are similar in construction to single-impact helmets in that they are composed of an outer shell designed to spread out the force of impact and engage as much of the energy-absorbing liner underneath during an impact. The energy-absorbing liner typically comprises either vinyl nitrile or expanded polypropylene foam. Vinyl nitrile foam is excellent at managing low-energy multiple impacts but tends not to perform as well for higher energy impacts. This type of foam returns to its original shape after an impact. Expanded polypropylene is similar to expanded polystyrene in its manufacturing process but has more elastic properties. As a result, the expanded polypropylene deforms under the impact, dissipating the impact energy returning to its original shape. As it can manage higher energies more effectively than vinyl nitrile foam, it is thought to be better for multi-impact applications [30] (Fig. 4.3).

Fig. 4.3 A snowboarding helmet is an example of multi-impact helmets

4.6.1.2 Mouthguard

Traumatic orofacial injuries commonly occur in fighting sports such as boxing or martial arts but also in contact sports like basketball, football, rugby and others [12]. Those kinds of injuries are often irreversible and can lead to functional, aesthetic and psychological problems.

The mouthguard acts by absorbing and dissipating a large portion of the energy in the impact zone.

According to the Academy for Sports Dentistry's (ASD) definition, three different types of mouthguards are available now [38]:

Type 1 is the stock type, which has no capability to adjust to an individual's morphological characteristics.
Type 2 is the "mouth formed" or "boil-and-bite" type. Both types 1 and 2 are commercially available and referred to as over-the-counter or on-the-shelf type.
Type 3 is the custom-made mouthguard including the single-layer and the multilayer laminated type.

In recent years, mouthguards have become a core component of the safety equipment used to reduce the risk of oral injury during athletic activities, with numerous studies suggesting their preventive effects against traumatic injury during sports [33, 35, 45].

Custom-made mouthguards are the most highly recommended for prevention of orofacial and dental injuries and are the best as regards retention, comfort, fit, resistance to tearing and ease of breathing [4].

4.6.1.3 Shoulder Pads

Shoulder pads are commonly used in high-contact sports like rugby or US football to protect shoulders during the tackle. Shoulder pads are designed to protect the shoulder and reduce the probability of injury by attenuating and dissipating the impact energy.

Shoulder pads did not reduce enough the force applied to the shoulder in the tackle technique of those players. Several researchers have suggested that the shoulder pads currently permitted do not prevent major musculoskeletal injury, e.g. dislocation and fracture [66], but they may prevent superficial injury around the shoulder complex, especially the acromioclavicular joint [26]. Harris and Spears concluded that shoulder pads were unlikely to prevent serious injuries. The reason of this can be found in the fact that players might tackle harder because they feel their shoulder pads would protect them from injury.

4.6.1.4 Wrist Guard

Wrist injuries are frequent in a large range of sports activities, notably snowboarding, skateboarding and skating.

Larger, predominantly epidemiological studies have shown, in analyses of over 20,000 individuals, that the frequency of wrist fractures is reduced, by up to 80 %, in snowboarders who wear wrist guards [23, 52]. However, some results have suggested a link between wrist guards and more complex proximal forearm, elbow or shoulder injuries [10, 23], although this has been refuted by other studies [50]. It may be that in high-energy falls some fractures will be sustained by the upper limb anyway.

Current wrist guards not only provide some impact protection by covering the impact area of the hand as demonstrated by biomechanical testing [31] but also they limit wrist movement, absorbing energy during the normal protective mechanism of falling on the outstretched extended wrist (Fig. 4.4).

Fig. 4.4 A model of wrist guard for snowboarding or skateboarding

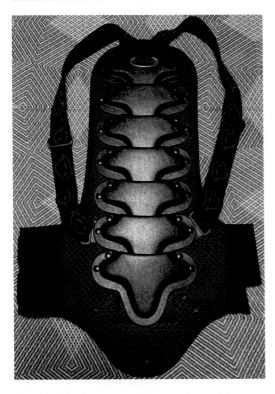

4.6.1.6 Shin Guards

Shin guards are frequently used, in football, hockey and skiing (especially in slalom). Their main function is to protect the soft tissues and bones in the lower extremities from external impact. Shin guards provide shock absorption and facilitate energy dissipation, thereby decreasing the risk of serious injuries. Currently, rigid materials (plastic, carbon, kevlar, etc.) are used for the outer shell, while soft materials are preferred as the lining of the guard. A well-designed shin guard should provide adequate protection for the shank, but allow range of motion of the ankle and the knee [20].

Many authors agree that shin guards used in football may reduce the number of minor injuries [3, 19]; however, it is unclear whether they can prevent more serious injuries such as tibia fractures. An in vitro study has demonstrated that the use of shin guards may not prevent fractures [2] (Fig. 4.6).

4.6.1.7 Ankle Braces/Taping

Taping and bracing of the ankle are common practices in the athletic setting. Generally, taping and bracing are used prophylactically in an effort to prevent a first-time ankle sprain or, more often, to prevent recurrent ankle sprains. Ankle orthotics is generally considered effective in providing mechanical stability while restricting joint range of motion. A significant amount of research on taping and bracing has been conducted; authors have investigated if these measures can reduce the incidence of lateral ankle sprains.

A recent review provides good evidence for the use of either ankle taping or ankle bracing to prevent lateral ankle sprains among previously injured players. However, for those without previous ankle injuries, this still needs to be proven. There is no evidence on which external ankle support is better than the other; there was no significant difference between ankle tape and ankle brace. Both can effectively reduce incidence of ankle sprains among previously injured individuals [15].

4.6.1.8 Shoe

Footwear has a significant effect on noncontact ligament injuries because it alters the fixation of

Fig. 4.5 A back protector with straps designed for snow sports

4.6.1.5 Back Protector

There are various back protectors available on the market, to prevent spine and back injuries. In recent years, these protectors have become very popular in snow sports and motorcycling.

There are two primary constructions found for such protectors on the market: an outer surface of hard shells, with padding underneath, and one or more layers of padding alone which can, for instance, be a viscoelastic foam. The protectors are typically held to the body with straps, with some models incorporated in clothing or backpacks.

Back protectors for snowboarders and skiers were analysed recently to determine their potential to prevent spinal injury [54], but no evaluation of their effectiveness was made.

Regarding the effectiveness of back protector for motorcyclist, a recent study shows that they are highly effective in reducing the probability of serious spinal injury (i.e. spinal fracture or spinal cord injury) [22] (Fig. 4.5).

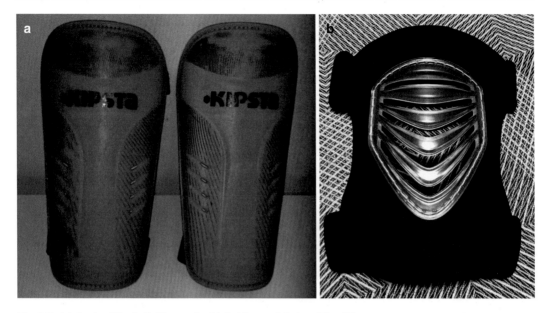

Fig. 4.6 (**a**) A pair of football shin guards. (**b**) A shin guard designed for skiing

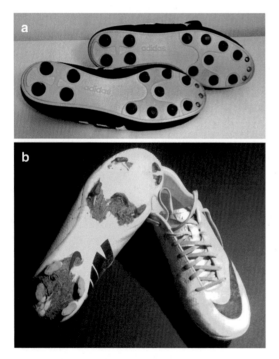

Fig. 4.7 (**a**, **b**) Football shoes with different numbers, distributions and lengths of the cleats

Cleated shoes vary in terms of number, diameter, length and placement of the cleats. Cleat styles vary among different sports, being optimised for the special needs of each sports gesture (Fig. 4.7).

Research has shown that the total number, tip diameter and length of the cleats may affect ACL injury rates because shoes with fewer, narrower and longer cleats result in an increased incidence of injury [63]. The reason that shoes with fewer, narrower and longer cleats caused a greater incidence of ACL injury was because these shoes created a higher coefficient of friction with the playing surface, especially in rotational movements [64].

4.6.2 Passive Protection

According to the definition of passive approach in preventive medicine by Haddon [74], passive protection from injury involves techniques to prevent injury and does not require any human action to be effective. These are automatic features designed to increase safety.

In terms of passive protection to prevent sports injury, playground's features and protection systems around the playing area play an important role.

the foot to the surface. Shoes with protruding cleats are the primary footwear used in sports associated with higher ACL injury rates [43].

4.6.2.1 Type of Surface

The playing surface has a significant effect on noncontact injuries because it affects the grip of the athlete's feet during movement.

For sports that may be played outdoors or indoors (such as football, Australian and US football), the two primary surfaces are natural grass and some variation of artificial turf. For sports that are only played indoors (such as basketball and team handball), the two primary surfaces are wood floor and artificial rubberized floor. In general, artificial turf has a higher coefficient of friction (COF) than natural grass [18, 69], and artificial rubberized floor has a higher COF than wood floor [44]. A higher COF is associated with an increased rate of ligament injury, especially of the ACL [43]. However, a surface with a too low COF may lead to slipping and falling injuries.

Based on previous research, athletes can therefore reduce their injury risk, thanks to the possibility of modifying the shoe–surface interaction by selecting their athletic footwear carefully, especially for what concerns cleat specifications according to the playing surface conditions.

Training programmes could be used to teach athletes to know the features of the different surfaces that they might encounter during the season.

4.6.2.2 Crash Measures

Ski

Different safety nets (A-nets, B-nets, triangular nets) are used in alpine ski to protect the racers when accidents occur. Spill zones (the area between the track and the safety nets) are also cleared to allow skiers to slide unobstructed if they fall [62].

The purpose of those safety devices is to absorb the kinetic energy of a falling skier, as well as to avoid them from crashing into danger areas (e.g. trees, pylons).

A systematic video analysis of 69 injury cases in 2009 World Cup alpine skiing showed that nets were very efficient in the majority of cases at stopping the skier, preventing a worse outcome of the fall [6].

Motor Sports

Around the race track, there must be no fixed objects and sand banks; tyre barriers or metal rails must be positioned around the circuit to absorb the kinetic energy of the driver and his vehicle during an off-track.

Replacing tyre barriers and metal rails with temporary crash protection barriers made of steel tubes and pads of hard foam may absorb some of the crash energy, reducing the loading to both head and neck during dramatic decelerations [37]. Crucial is the wideness between the circuit and the barriers that allows the driver and the vehicle to lose speed before impact.

References

1. Anderson B, Burke ER (1991) Scientific, medical, and practical aspects of stretching. Clin Sports Med 10:63–86
2. Ankrah S, Mills NJ (2003) Performance of football shin guards for direct stud impacts. Sport Eng 6:207–220
3. Arnason A et al (2004) Risk factors for injuries in football. Am J Sports Med 32(1 Suppl):5S–16S
4. Badel T, Jerolimov V, Panduric J et al (2007) Custom-made mouthguards an prevention of orofacial injuries in sports. Acta Med Croatica 61(Suppl 1):9–14, Croatian
5. Barengo NC et al (2014) The impact of the FIFA 11+ training program on injury prevention in football players: a systematic review. Int J Environ Res Public Health 11(11):11986–12000
6. Bere T et al (2013) A systematic video analysis of 69 injury cases in World Cup alpine skiing. Scand J Med Sci Sports 24(4):667–677
7. Biasca N, Wirth S, Tegner Y (2002) The avoidability of head and neck injuries in ice hockey: an historical review. Br J Sports Med 36(6):410–427, Review
8. Brown C (2005) Injuries: the psychology of recovery and rehab. In: Murphy S (ed) The sport psych handbook. Human Kinetics, Champaign, pp 215–235
9. Butterfield GE (1987) Whole body protein utilization in humans. Med Sci Sports Exerc 19:S157–S165
10. Chow TK, Corbett SW, Farstad DJ (1996) Spectrum of injuries from snowboarding. J Trauma 41(2):321–325
11. Cirelli C, Tononi G (2008) Is sleep essential? PLoS Biol 6(8):e216
12. Cohenca N, Roges RA, Roges R (2007) The incidence and severity of dental trauma in intercollegiate athletes. J Am Dent Assoc 138(8):1121–1126

13. Cornelius WL (1983) Stretch evoked EMG activity by isometric contraction and submaximal concentric contraction. Athl Train 1983(18):106–109

14. Coyle EF, Hagberg JM, Hurley BF et al (1983) Carbohydrate feeding during prolonged strenuous exercise can delay fatigue. J Appl Physiol 55:230–235

15. Dizon JM, Reyes JJ (2010) A systematic review on the effectiveness of external ankle supports in the prevention of inversion ankle sprains among elite and recreational players. J Sci Med Sport 13(3):309–317

16. Dodd S, Powers SK, Callender T et al (1984) Blood lactate disappearance at various intensities of recovery exercise. J Appl Physiol 57:1462–1465

17. Dosil J (2004) Psicología de la actividad física y del deporte. McGraw Hill, Madrid

18. Dragoo JL, Braun HJ (2010) The effect of playing surface on injury rate: a review of the current literature. Sports Med 40(11):981–990

19. Ekstrand J, Gillquist J (1982) The frequency of muscle tightness and injuries in soccer players. Am J Sports Med 10(2):75–78

20. Eugene B (2003) The prevention of injuries in youth soccer. Michigan Governor's Council on physical fitness, health and sports, November. Available form: http://files.leagueathletics.com/Images/Club/6097/The%20Prevention%20of%20injuries.pdf

21. European Food Information Council (2009) Food-based dietary guidelines in Europe. http://www.eufic.org/article/en/expid/food-based-dietary-guidelines-in-europe/. Cited 15 Dec 2013

22. Giustini M, Cedri S, Tallon M et al (2014) Use of back protector device on motorcycles and mopeds in Italy. Int J Epidemiol 43(6):1921–1928

23. Hagel B, Pless IB, Goulet C (2005) The effect of wrist guard use on upper-extremity injuries in snowboarders. Am J Epidemiol 162(2):149–156

24. Halson SL (2014) Sleep in elite athletes and nutritional interventions to enhance sleep. Sports Med 44(Suppl 1):S13–S23, Review

25. Hardy J, Crace K (1980) Dealing with injury. Sport Psychol Training Bull 6:1–8

26. Harris DA, Spears IR (2008) The effect of rugby shoulder padding on peak impact force attenuation. Br J Sports Med 44(3):200–203

27. Hedrick A (1992) Physiological responses to warm-up. Natl Strength Cond Assoc J 14(5):25–27

28. Hewett TE, Myer GD, Ford KR, Slauterbeck JL (2006) Preparticipation physical exam using a box drop vertical jump test in young athletes: the effects of puberty and sex. Clin J Sports Med 16:298–304

29. Hoshizaki TB, Brien SE (2004) The science and design of head protection in sport. Neurosurgery 55(4):956–966, discussion 966–967

30. Hoshizaki TB, Post A, Oeur RA et al (2014) Current and future concepts in helmet and sports injury prevention. Neurosurgery 75(Suppl 4):S136–S148

31. Hwang IK, Kim KJ (2004) Shock-absorbing effects of various padding conditions in improving efficacy of wrist guards. J Sports Sci Med 3(1):23–29

32. Ivy JL, Datz AL, Cutler CL et al (1988) Muscle glycogen synthesis after exercise effect of time of carbohydrate ingestion. J Appl Physiol 64:1480–1485

33. Johnsen DC, Winters JE (1991) Prevention of intraoral trauma in sports. Dent Clin North Am 35:657–666

34. Kay AD, Blazevich AJ (2012) Effect of acute static stretch on maximal muscle performance: a systematic review. Med Sci Sports Exerc 44(1):154–164

35. Knapik JJ, Marshall SW, Lee RB et al (2007). Mouthguards in sport activities: history, physical properties and injury prevention effectiveness. Sports Med 37:117–144 [3–6, 9, 10, 14, 19, 21, 22]

36. Konrad A, Gad M, Tilp M (2015) Effect of PNF stretching training on the properties of human muscle and tendon structures. Scand J Med Sci Sports 25(3):346–355. doi: 10.1111/sms.12228

37. Lippi G, Guidi GC (2005) Effective measures to improve driver safety. Br J Sports Med 39(9):686

38. Maeda Y, Kumamoto D, Yagi K et al (2009) Effectiveness and fabrication of mouthguards. Dent Traumatol 25(6):556–564

39. Mah CD, Mah KE, Kezirian EJ et al (2011) The effects of sleep extension on the athletic performance of collegiate basketball players. Sleep 34:943–950

40. Malina R, Bouchard C, Bar-Or O (2004) Growth, maturation and physical activity. Human Kinetics, Champaign

41. McIntosh AS, Andersen TE, Bahr R et al (2011) Sports helmets now and in the future. Br J Sports Med 45(16):1258–1265, Review

42. Meredith CN, Zachin MJ, Fontera WR et al (1981) Dietary protein requirements and body protein metabolism in endurance trained men. J Appl Physiol 66:2850–2856

43. Noyes FR, Barber-Westin SD (2012) ACL injuries in the female athlete, 85. doi:10.1007/978-3-642-32592-2_4, © Springer-Verlag Berlin Heidelberg

44. Olsen OE, Myklebust G, Engebretsen L et al (2003) Relationship between floor type and risk of ACL injury in team handball. Scand J Med Sci Sports 13(5):299–304

45. Onyeaso CO (2004) Secondary school athletes: a study of mouthguards. J Natl Med Assoc 96:240–245

46. Palmi J (2002) Aspectos psicosociales en la prevención y recuperación de lesiones deportivas. In: Rodríguez L, Gusi N (eds) Manual de prevención e rehabilitación de lesiones deportivas. Síntesis, Madrid

47. Petitpas A, Danish S (1995) Caring for injured athletes. In: Murphy S (ed) Sport psychology interventions. Human Kinetics, Champaign, pp 255–281

48. Postolache TT, Oren DA (2005) Circadian phase shifting, alerting, and antidepressant effects of bright light treatment. Clin Sports Med 24(2):381–413

49. Reilly T, Edwards B (2007) Altered sleep–wake cycles and physical performance in athletes. Physiol Behav 90(2–3):274–284

50. Rønning R, Rønning I, Gerner T et al (2001) The efficacy of wrist protectors in preventing snowboarding injuries. Am J Sports Med 29(5):581–585

51. Rowbottom DJ (2000) "Periodization of training". In: Garrett, WE, Kirkendall DT (eds) Periodization of training. Lippincott Williams & Wilkins, Philadelphia, p 499. ISBN. Retrieved April 20, 2013

52. Russell K, Hagel B, Francescutti LH (2007) The effect of wrist guards on wrist and arm injuries among snowboarders: a systematic review. Clin J Sport Med 17(2):145–150, Review

53. Safran MR, Seaber AV, Garrett WE (1989) Warm-up and muscular injury prevention – an update. Sports Med 8(4):239–249

54. Schmitt KU, Liechti B, Michel FI et al (2010) Are current back protectors suitable to prevent spinal injury in recreational snowboarders? Br J Sports Med 44:822–826

55. Shellock FG, Prentice WE (1985) Warming-up and stretching for improved physical performance and prevention of sports-related injuries. Sports Med 2:267–278

56. Sherman WM, Brodowicz G, Wright DA et al (1989) Effects of 4-hour pre-exercise carbohydrate feedings on cycling performance. Med Sci Sports Exerc 2:598–604

57. Silvério J, Srebro S (2012) Como ganhar usando a cabeça. Afrontamento, Porto

58. Simonsen JC, Sherman WM, Lamb DL et al (1991) Dietary carbohydrate, muscle glycogen, and power output during rowing training. J Appl Physiol 70(4):1500–1505

59. Slavin J (1991) Protein needs for athletes. The health professional's handbook. Aspen Publisher, Gaithersberg

60. Spiegel K, Leproult R, Van Cauter E (1999) Impact of sleep debt on metabolic and endocrine function. Lancet 354(9188):1435–1439

61. Tarnopolsky MA, MacDougall JD, Atkinson SA (1988) Influence of protein intake and training status on nitrogen balance and lean body mass. J Appl Physiol 64:187–193

62. The International Ski Federation (FIS) (2011a) FIS safety material. http://www.fis-ski.com/data/document/simalist11-07-11.pdf

63. Torg JS, Quedenfeld T (1971) Effect of shoe type and cleat length on incidence and severity of knee injuries among high school football players. Res Q 42(2):203–211

64. Torg JS, Quedenfeld TC, Landau S (1974) The shoe surface interface and its relationship to football knee injuries. J Sports Med 2(5):261–269

65. Udry E, Gould D, Bridges D et al (1997) Down but not out: athlete responses to season-ending ski injuries. J Sport Exerc Psychol 3:229–248

66. Usman J, McIntosh AS, Fréchède B (2011) An investigation of shoulder forces in active shoulder tackles in rugby union football. J Sci Med Sport 14(6):547–552

67. Waterhouse J, Atkinson G, Edwards B et al (2007) The role of a short post-lunch nap in improving cognitive, motor, and sprint performance in participants with partial sleep deprivation. J Sports Sci 25(14):1557–1566

68. Weinberg R, Gould D (2014) Foundations of sport and exercise psychology. Human Kinetics, Champaign

69. Williams S, Hume PA, Kara S (2011) A review of football injuries on third and fourth generation artificial turfs compared with natural turf. Sports Med 41(11):903–923

70. Yesalis CE (1993) Anabolic steroids in sports and exercise. Human Kinetics, Champaign

71. Zawadski KM, Yaspelkis BB, Ivy JL (1992) Carbohydrate protein complex increases the rate of muscle glycogen storage after exercise. J Appl Physiol 72(5):1854–1859

72. Fradkin AJ, Gabbe BJ, Cameron PA (2006) Does warming up prevent injury in sport? The evidence from randomised controlled trials? J Sci Med Sport 9(3):214–220. Epub 2006 May 6. Review

73. Bandy WD, Irion JM, Briggler M (1997) The effect of time and frequency of static stretching on flexibility of the hamstring muscles. Phys Ther 77(10):1090–1096

74. Haddon W Jr. (1974) Editorial: Strategy in preventive medicine: passive vs. active approaches to reducing human wastage. J Trauma 14(4):353–354

General Prevention Principles of Overload Damage in Sports

5

Henrique Jones

Key Points

1. Overload and injury in sports.
2. Overtraining and biologic, immunological, and biochemical changes.
3. Mechanical overloading and articular degenerative changes.
4. Neuromuscular training programmes.
5. Analysis of Physical condition in sports.
6. The workout concept.
   Sports injury prevention including protective equipment.

5.1 Introduction

5.1.1 Definition of Overload

The principle of overload states that a greater than normal stress or load on the body is required for training adaptation to take place. The body will adapt to this stimulus. Once the body has adapted, then a different stimulus is required to continue the change. In order for a muscle (including the heart) to increase strength, it must be gradually stressed by working against a load greater than it is used to. To increase endurance, muscles must work for a longer period of time than they are used to. If this stress is removed or decreased, there will be a decrease in that particular component of fitness. A normal amount of exercise will maintain the current fitness level.

5.2 Coordination of Expectations and Performance

Athletes often set expectations on how they will perform in competition. These expectations could be based on their past results, competition factors, self-condition, environment and, in many cases, what they hear from other people, coaches or media.

Research has consistently shown that expectations often cause anxiety, especially when the athlete feels that his or her skills do not match up with the expected results. This can lead to an increase in precompetition and competition anxiety, often causing poor performance, and over time may even cause burnout. Many coaches and parents are unknowingly at fault when it comes to producing anxiety amongst athletes. The success fear syndrome and the failure fear syndrome are some examples of anxiety status becoming psychologic disturbance.

H. Jones, MD, MOS, MSM
Montijo Orthopedic and
Sports Medicine Clinic, Lusofona University,
Rua Miguel Pais, n° 45, 2870-356 Montijo,
Lisbon, Portugal
e-mail: ortojones@gmail.com

© ESSKA 2016
H.O. Mayr, S. Zaffagnini (eds.), *Prevention of Injuries and Overuse in Sports: Directory for Physicians, Physiotherapists, Sport Scientists and Coaches*, DOI 10.1007/978-3-662-47706-9_5

To improve performance, the athlete must first build confidence and focus on the things that they have the most control over. If the athlete has low confidence, then he, or she, should focus on the poorest results that they feel they can do well, even if it is to simply smile or go out and compete.

Confidence is mostly produced by the results an athlete receives. It is no wonder that setting small goals helps athletes produce more confidence because the athlete can see and experience more positive results.

Expectations and goals are always present in the beginning of sports season, and with that, along with their new year's resolutions, many athletes begin to set goals for the upcoming season. When someone knows how to properly use goals and expectations, this could be an enormous help to cut through obstacles, overcome blocks, motivate to get back up after falling and, ultimately, carve athletic dream into a wonderful reality. However, when someone misuses goals and expectations, bringing them out at the wrong time, physical and mental disturbances can occur with decrease of performance. Athletes who misuse their goals continually experience frustration when they compete because they just cannot seem to execute the way that they are capable. Furthermore, the pressure that they experience right before and during competition makes it impossible for them to relax and have fun. Without the ability to feel loose and enjoy themselves, these athletes are more prone to performance slumps, fears, blocks and failure. If reaching your dream as an athlete is vitally important to you, if you want your son or daughter to really enjoy their sports experience and be as successful as possible and if you, as a coach, want your athletes to come through in the clutch and consistently perform like champions, then it is critically important for you to understand the simple bases of goals and expectations.

5.3   Analysis of Laboratory Parameters

Many laboratory changes can occur after exercise and recover. Serum ferritin concentrations can be depressed significantly from pretraining concentrations at the conclusion of the recovery period, while the expression of lymphocyte activation antigens (CD25+ and HLA-DR+) can be increased, both after the training phase and the recovery phase. The number of CD56+ cells in the peripheral circulation can be depressed at the conclusion of the recovery period. Several parameters can change in association with overload training failing to reflect the decrease in performance experienced by athletes, suggesting that overtraining may best be diagnosed through a multifactorial approach to the recognition of symptoms.

The most important factor to consider may be a decrease in the level of performance following a regeneration period. The magnitude of this decreased performance necessary for the diagnosis of overtraining and the nature of an "appropriate" regeneration period are, however, difficult to define and may vary depending upon the training background of the subjects and the nature of the preceding training. It may be associated with biochemical, haematological, physiological and immunological indicators, but many doubts still exist. Individual cases may present a different range of symptoms, and diagnosis of overtraining should not be excluded based on the failure of blood parameters to demonstrate variation. However, blood parameters may be useful to identify possible aetiology in each separate case report of overtraining.

5.3.1   Possible Immunological and Biochemical Markers of Impending Overtraining

It is difficult to identify, and correlate, overtraining, fatigue and biomarkers. Individual metabolicresponses to exercise are very different regarding

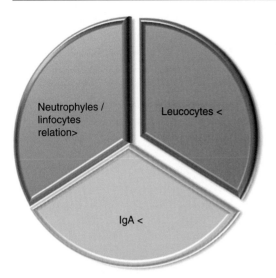

Fig. 5.1 Most common laboratory changes after intensive exercise

several factors as training condition, skills, gender, recovery procedures, diet and hydration, amongst others. It is universally accepted that the relationship between neutrophils and lymphocytes [1] is alternated after exercise, as well as the decrease of leucocytes and immunoglobulin A (Figs. 5.1 and 5.2). However, there are other laboratory parameters that can show the biologic modifications happening during and after exercise (Table 5.1), which can be monitored in a battery of tests (Table 5.2).

5.3.2   Omegawave Test: One More Tool to Understand Fatigue and Overtraining

Omegawave device can help, in association with other factors, identify limiting factors for CNS and cardiac and metabolic systems. Omegawave's

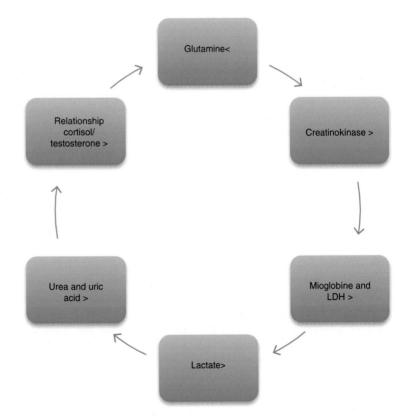

Fig. 5.2 Blood parameters and its modifications after vigorous exercise

Table 5.1 Immunological and biochemical markers associated with overtraining

Leucocyte responses to antigens (e.g. lymphocyte proliferation, neutrophil degranulation, NK cytotoxic activity)
Salivary IgA
Neutrophil/lymphocyte ratio• T-cell CD4+/CD8+ Ratio• T-cell CD4+CD45RO+ expression
Plasma cortisol or cortisol/testosterone ratio
Urinary steroids or catecholamines
Plasma glutamine
Plasma urea
Plasma cytokines (e.g. IL-6)
Blood lactate response to incremental or high-intensity exercise
Plasma or salivary cortisol response to high-intensity exercise

Table 5.2 Individual monitoring: a suggested battery of tests to detect impending overtraining

Performance
Mood state questionnaires
Diary of responses to training (fatigue, muscle soreness) and symptoms of illness
Sleeping heart rate
Blood lactate and plasma cortisol response to high-intensity or incremental exercise
Plasma creatine kinase activity
Cortisol to testosterone ratio
Nocturnal urinary noradrenaline and adrenaline secretion
Routine haematology (blood haemoglobin, serum ferritin, leucocyte counts)
T-lymphocyte CD4+/CD45RO+ expression
Experience of coach and athlete

measures can be adjusted to validate training approach and assess the functional readiness of athletes, with the aim of identifying the optimal types and intensities of training and recovery, to improve athletic performance and help avoid injury. It takes measurements relevant to an athletes' physiological condition, including ECG, Omega (DC potential of the brain), neuromuscular and reaction rate measurements, for analysis. Measurements and their results can be taken for a whole team and viewed by a coach locally or remotely. The measurements are processed by Omegawave's patented cloud-based system to give results and recommendations that the company claims are relevant to the athlete's cardiac readiness (Fig. 5.3), metabolic readiness, central nervous system readiness, gas exchange readiness, detoxification readiness and hormonal system readiness (Fig. 5.4) in the so-called omegametry (electrophysiological padronised values that provide information about physical surcharge).

Rooted in Russian sports science and space medicine, Omegawave technology was developed in the USA. Its patented technology was originally based on physiological data from 10,000 high-performing athletes and has been validated in countless studies conducted by sports and research institutes worldwide.

Unlike the traditional stress test, known as the VO$_2$Max or treadmill test, the Omegawave system takes readings of athletes during times of rest or relaxation. The idea is to measure how quickly athletes are recovering from workouts, competition and other stress [2].

Based on regular readings of the tests, coaches and trainers can adjust training regimens, tweaking the cycle of stress and recovery in athletes to maximise performance.

The test begins with the athlete sitting or lying down, electrodes attached from the body (Fig. 5.5) to a device that connects to a laptop computer. Within 5 min, the device generates results for the systems that regulate cardiac activity and metabolism. A test that produces results for the central nervous, hormonal and detoxification systems takes 10 more minutes.

The system measures brainwave activity to determine how the athlete is adapting to stress of all kinds. It might find that an athlete is recovering quickly and can handle a more intense regimen. Or it might indicate that the body is being overly taxed and needs more rest.

5.3.3 Is Mechanical Overloading Also a Problem?

Under high force and long duration loading, the collagen fibres exhibited high deformation with an increased thickness of the layer of collagen

Fig. 5.3 Normal presentation of cardiac evaluation by Omegawave device

Date of birth: February 26, 1983 (27 years old)
Weight, Height: 79 kg. 187 cm.

ECG

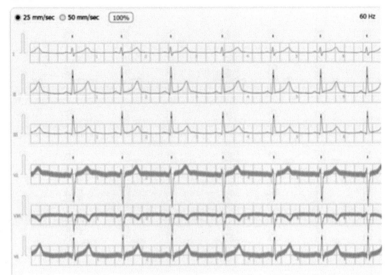

Weight, Height: 79 kg. 187 cm.

Athlete Readiness

Based on the HRV assessment:
Cardiac system is reasonably ready for any level of activity.

Based on the Omega assessment:
CNS: Sufficient resistance to physical and psychological stress.

Athlete Readiness Overall

Readiness Indicators

Autonomic Function

Current state of Cardiac System		
Stress index	6	Within the norm
Fatigue	7	Complete recovery
Adaptation reserves	5	Moderate

Current state of Regulatory Mechanisms		
CNS	4	Normal
GEC System	-	-
Detoxification System	-	-
Hormonal System	-	-

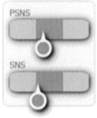

PSNS

SNS

Current state of Energy Supply System				
Parameter	Grade	Value	Norm	
Aerobic status index	4	111	110 - 160	Moderate
Anaerobic status index	6	132	132 - 160	Low
Alactic status index	4	13	12 - 25	Moderate

Grades 1-7, 7 is optimal

Fig. 5.4 Normal tracing of athlete readiness in Omegawave device

Fig. 5.5 Athlete during Omegawave testing

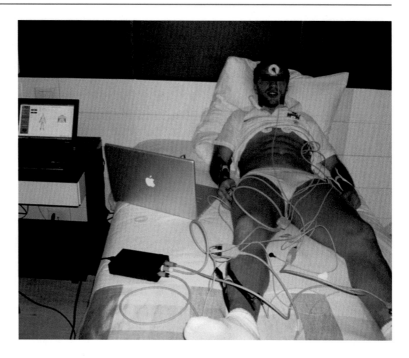

fibres oriented almost parallel to the surface and a cartilage thickness reduced to 54 % [3]. Both overuse and disuse of joints upregulate matrix metalloproteinases (MMPs) in articular cartilage and cause tissue degradation; however, moderate (physiological) loading maintains cartilage integrity [4, 5]. In top athletes, joints surcharge and cartilage degradation can be a handicap, regarding performance, through the years. Besides genetics, load type, frequency and time of exposure, the decrease of proteoglycans and type II collagen and the increase of type I collagen metalloproteinases, isoleucines and nitric oxide can explain the chondrocyte failure and the precocity cartilage degradation in top-level athletes (Fig. 5.6).

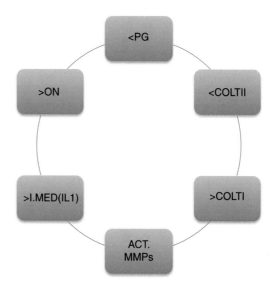

Fig. 5.6 Metabolic joint cartilage changes, in top athlete, in relationship with continuous and vigorous exercise

5.4 Analysis of the Movement Sequence and Neuromuscular Training Programmes

Modern human movement analysis is the interpretation of computerised data that documents an individual's upper and lower extremities, pelvis and trunk motion during movement. The development and improvement of electronic technology,

and computer science, has made it easier to analyse human movement [6].

The increasing involvement of technology in medicine has made some other methods and tools available for tracking and assessing human motion. Additional equipments, such as dynamic EMG, force plates, pedobarographs, electrogoniometers

Fig. 5.7 Analysis of movement in laboratory

and metabolic analyzers, have made a more complicated but also more complete acquisition of human movement available. Sports performance is directly linked to human motion and performance. So, movement analysis is automatically a part of human performance assessment and analysis.

Today in many sports, sports scientists use movement analysis as a tool to enhance techniques, correct movement errors, assess metabolic costs related to a variety of movements and aid in rehabilitation and prevention.

The deleterious impact of injuries on athlete health and performance has precipitated extensive efforts towards their prevention. In particular, recent research has focused on identifying underlying neuromuscular control predictors, as such factors are readily amenable to training and in essence, preventable, after biomechanic laboratory studies (Fig. 5.7).

Neuromuscular training programmes continue to evolve out of this research, which attempt to

modify what are considered abnormal and potentially hazardous movement strategies. This in turn would facilitate more effective prevention methods that promote successful neuromuscular adaptation within individual non-modifiable constraints, like individual neuromuscular fatigue profile.

According to the work of Hubsche et al. [7] based on the results of seven high-quality studies, to assess the effectiveness of proprioceptive/neuromuscular training in preventing sports injuries, it seems evident that proprioceptive/neuromuscular training is effective in reducing the incidence of certain types of sports injuries amongst adolescent and young adult athletes during pivoting sports.

5.5   Analysis of Physical Condition

It is widely known that a scientific approach has to be applied for selecting potential athletes and training them for better performance. As a fundamental step for adopting a scientific approach, systematic collection of empirical data and materials are essential. Until now, many sports scientists, in cooperation with sports doctors and physiotherapists, have conducted diverse kinds of research to elucidate various characteristics of elite athletes including morphology, fitness, psychology, behaviour, skills and physical capacity. The majority of these studies were based on the scientific analyses of elite athlete's physical characteristic fitness status or competition records as shown in Melnikov and Singer [8] work.

The rationale for continuous endeavour of the scientists studying elite athletes who are in the middle of their peak performance stage is in the fundamental hypothesis that a specific mode of athletes' performance is ideally optimised by their physical characteristics and fitness as well as their psychological status and techniques.

The concept of elite athlete as model of the sports events, in which they are involved, could be related as a profile and a contribution to activities as selecting, training and prescribing athletes, having in mind that physical characteristics such as body morphology and fitness are relatively simple to quantify, and reliable to measure, compared to the other functional parameters, according to Yang and Lee [9].

Normally, a general and major attempt to formulate and clarify physical fitness profiles of the elite athletes was describing and comparing means and standard deviations of each measured variable of a specific event. And analyses of the differences between means were also utilised. However, this profile information may only be useful to compare characteristics amongst athletes of various sport disciplines but have a limitation in providing training guidelines and prescriptive measures for those who are in a beginner's and developmental stages. To be used as an informative tool applicable for selection of young athletes and training of elite athletes, a profile which can be adoptable for every age group should be formulated. Thus, the standard of comparison for each parameter should be formed by the values which are elicited from the general population that are not elite athletes and are age-matched general counterparts, according to Matsudo [10].

In other words, the degree of superiority of elite athletes of each sport event should be directly compared with the general population of nonathletes in terms of each measured variable and their profiles. This may be possible while, based on the mean value of the general population, the mean value of the elite athlete of each sport event is converted to z value or T-score by utilising Standard Difference Analysis (SDA). Ball games including football, basketball, baseball and volleyball are popular and well-known sport events. These ball games have their own professional leagues, and the players of these games are thought to have very promising professions. Along the same line, children, adolescents and their parents are interested in finding necessary morphological characteristics and fitness levels for performing a specific sport according with previous studies, programmes and guidelines.

The study of Byoung and Kim [11] aimed to construct physical and fitness profiles of elite ball game players in comparison to the age-matched general population and, based on the data, to establish a training evaluation index applicable for athletes of various age ranges and to use the index as a tool for selecting young ball players. For elite football players, the evaluation reveals a profile of high-seated height, lower body fat contents, high levels of upper body muscular strength and endurance, upper body power, abdominal muscular endurance, lower body muscular strength and endurance, lower body power, cardiorespiratory endurance, speed, agility and flexibility. For elite basketball players, the profile reveals the superiority of height, seated height, body weight and chest circumference as well as high levels of upper body power, abdominal muscular endurance, lower body lower body power, cardiorespiratory endurance, speed, flexibility and agility.

Despite the importance of sportsmen, and sportswomen, characterisation and selection, regarding physical and morphologic qualities, for sports participation, the physical demands in sports have been studied intensively, and the knowledge of metabolic changes during a game and their relation to the development of fatigue have been studied per criteria [12].

Heart rate and body temperature measurements suggest that for elite sports participation, the average oxygen uptake during a competition, like a football match, is around 70 % of maximum oxygen uptake (VO2max). A top-class sportsman has 150–250 brief intense actions during a game, indicating that the rates of creatine-phosphate (CP) utilisation and glycolysis are frequently high during a game, which is supported by findings of reduced muscle CP levels and severalfold increases in blood and muscle lactate concentrations. Likewise, muscle pH is lowered and muscle inosine monophosphate (IMP) elevated during a football game. Fatigue appears to occur temporarily during a competition, but it is not likely to be caused by elevated muscle lactate, lowered muscle pH or change in muscle-energy status. It is unclear what causes the transient reduced ability of sportsmen to perform maximally. Muscle glycogen is reduced by 40–90 % and is probably the most important substrate for energy production, and fatigue towards the end of a game might be related to depletion of glycogen in some muscle fibres. Blood glucose and catecholamines are elevated and insulin lowered during a game. The blood free-fatty-acid levels increase progressively, probably reflecting an increasing fat oxidation compensating for the lowering of muscle glycogen. Thus, elite athletes, in mixed sports competitions, like football, have high aerobic requirements throughout a game and extensive anaerobic demands during periods of a match leading to major metabolic changes, which might contribute to the observed development of fatigue during and towards the end of the competition. The analysis of the variants exposed above could play an important role in the evaluation of an athlete's physical condition, and readiness, regarding next competitions.

A blood sample collected after the competition, and the day after, to analyse the CP concentration in blood could be an easy way to monitor fatigue, and recover, in top competition athletes, associated with urine tests (Fig. 5.8) to screen dehydration and abnormal contents like proteins, urobilinogen, blood or leucocytes.

5.6   The Workout Concept

For any individual who is physically active, there is a possibility of sustaining an injury. While some injuries are unpredictable, such as an ankle

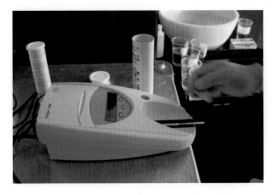

Fig. 5.8 Urine analyser, a simple way to collect athlete physical condition data

sprain or fracture, and difficult to prevent, many other injuries are preventable. By following a few simple guidelines, injuries such as muscle strains, tendinopathies and overuse injuries can be reduced.

The traditional training phases associate warm-up, progressivity, plasticity, volume, intensity, alternation, continuity, motivation and cooldown. The modern concepts on sports tend to congregate these aspects on a continuous sports specificity work with integrated ball training sessions and preventing exercises.

Every workout should begin with a warm-up and end with a cooldown. A warm-up is necessary to prepare the body for exercise by increasing heart rate and blood flow to working muscles. The warm-up (Fig. 5.9) should start slow and easy and consist of a general cardiovascular exercise such as walking, jogging or biking. After 5–10 min (40 % of maximum heart rate), the warm-up should focus on muscles and movements more specific to the exercise activity planned.

The second part of a warm-up regimen, to be performed immediately after the aerobic warm-up and as soon as possible before a practice or match, involves dynamic stretching (stretching muscles while moving). For sports that involve rapid movement in different directions, such as football, basketball, volleyball and tennis, players need to perform stretching exercises that involve many different parts of the body, by creating a smooth transition from the warm-up to a specific activity. For example, a football player could pass, dribble and shoot a ball; a weightlifter could lift lightweights before moving onto greater resistance.

Flexibility is absolutely a part of every good warm-up. Once the muscles are warm, they become more elastic and are ready to be stretched. Whether you choose to perform static stretches (by holding each position for 10–30 s) or perform dynamic stretches (by moving the body through a functional range of motion), flexibility prepares the muscles, tendons and joints for work by allowing them to move freely through a full-active range of motion. It is generally accepted that the more prepared the body is, the less likely it is to get injured and that increasing the flexibil-

Fig. 5.9 Warm-up in
football

Fig. 5.9 Warm-up in
football

Fig. 5.10 Stretching exercises before training

ity of a muscle-tendon unit promotes better performances and decreases the number of injuries.

Stretching exercises are regularly included in warm-up and cooling-down exercises; however, contradictory findings have been reported in the literature [13]. Several authors have suggested that stretching (Fig. 5.10) has a beneficial effect on injury prevention. In contrast, clinical evidence suggesting that stretching before exercise does not prevent injuries has also been reported. Apparently, no scientifically based prescription for stretching exercises exists, and no conclusive statements can be made about the relationship of stretching and athletic injuries. Stretching recommendations are clouded by misconceptions and conflicting research reports and part of these contradictions can be explained by considering the type of sports activity in which an individual is participating.

Sports involving bouncing and jumping activities with a high intensity of stretch-shortening cycles (e.g. soccer and football) require a muscle-tendon unit that is compliant enough to store and release the high amount of elastic energy that benefits performance in such sports. If the participants of these sports have an insufficient compliant muscle-tendon unit, the demands in energy absorption and release may rapidly exceed the capacity of the muscle-tendon unit. This may lead to an increased risk for injury of this structure. Consequently, the rationale for injury prevention in these sports is to increase the compliance of the muscle-tendon unit. Some studies [14] have shown that stretching programmes can significantly influence the viscosity of the tendon and make it significantly more compliant, and in case of sport demands of high intensity, stretching may be important for injury prevention. This conjecture is in agreement with the available scientific clinical evidence from these types of sports activities.

In contrast, when the type of sports activity contains low-intensity or limited stretch-shortening cycles (e.g. jogging, cycling and swimming), there is no need for a very compliant muscle-tendon unit since most of its power generation is a consequence of active (contractile) muscle work that needs to be directly transferred (by the tendon) to the articular system to generate motion. Therefore, stretching (and thus making the tendon more compliant) may not be advantageous. This conjecture is supported by the literature [13], where strong evidence exists that stretching has no beneficial effect on injury prevention in these sports. If this point of view is used when examining research findings concerning stretching and injuries, the reasons for the contrasting findings in the literature are in many instances resolved.

An area that often gets ignored is the cooldown after activity. Just as the warm-up prepares the body for work, the cooldown brings it back to its normal state. Time spent performing 5–10 min of low-intensity cardiovascular activity followed by stretching immediately after the workout will decrease muscle soreness and aid in recovery, both helping to prepare the body for the next workout.

For those people that enjoy sports, in a non-professional level, once an exercise programme is developed, there are a few things to remember. Start slow: people often jump right into a workout and do too much too fast, creating excessive muscle soreness and tightness. Proper progression is the key to preventing injuries. Slowly increase the amount of time of each workout, the intensity of the workout and the resistance of the weights. A 5 % increase as the exercise becomes too easy is a safe progression. Exercise at a level that is appropriate for your age and your fitness level. A young athlete competing with older children may not be as physically strong, predisposing them to injury. The same can be true for a weekend warrior athlete who jumps into a game with athletes who have trained throughout the week. If equipment is involved in your exercise programme, take the time to ensure that you have the proper equipment, that it fits correctly and that it meets safety standards. Too often, old, faulty

or improperly fitted equipment, such as footwear, mouthguards, helmets, goggles or shin pads, can cause injuries.

One of the best ways to prevent injury is to listen to the warning signs of the body. By ignoring little aches and pains in joints and muscles, a more serious injury could develop. Pain is the body's way to express that something is not right! The common expression "no pain, no gain" creates a large misconception. It is very possible to make cardiovascular and strength gains in workout routine without causing pain.

Rest is a critical component to any good workout routine and time spent allowing the body to recover is a great way to prevent injuries. A rest day must occur at least one to two times per week. Even small breaks during a workout are sometimes required to get the most out of the workout and prevent injuries.

A healthy, well-balanced diet can aid in injury prevention as well. A poor diet can lead to muscle weakness, decreased muscle strength and endurance. Equally important is maintaining hydration throughout the day, during and after your workout. A body with adequate fuel (food and water) will stay sharp and keep moving at the intensity you desire.

5.7   Passive and Active Protection

5.7.1   Sports Protective Equipment and Sports Injury Prevention

As a result of injuries to athletes during sporting activities, safety standards are set by government, national health and public health organisations to identify risks and protective equipment required in specific sports, particularly action or high contact sports, to reduce risk of injury. Athletes that compete professionally or as part of employment are protected under occupational safety and health standards. Protective equipment [15] may include helmets, protective eyewear, mouthguards, face protection, shin guards, jock straps, life jackets, safety mats, pads and guards, protective footwear and

Fig. 5.11 An example of football player's helmet

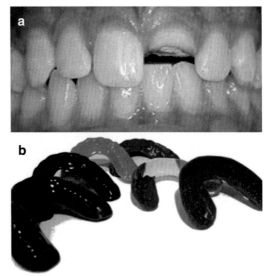

Fig. 5.12 (**a**) Tooth fracture after direct impact and (**b**) mouthguards

padded flame-resistant pressure suits for motorcyclists and motocross participants. This is particularly important when the sport, or activity, involves physical contact with other players and participants.

5.7.1.1 Helmets

To prevent or minimise head or brain injuries to sports people, such as boxers, cricketers, football players (Fig. 5.11), cyclists, skiers, baseball and motor sports, helmets are mandatory or recommended. These helmets are specially designed and tested according to the impacts of different types of sports, so a baseball helmet cannot be worn by a cyclist or boxer. The helmet should fit the player's head properly to also prevent damage from wear.

5.7.1.2 Protective Eyewear

Protective sports eyewear may include specially designed sunglasses for skiers or snowboarders, such as 3-mm polycarbonate lenses with ultraviolet filter to protect the eyes from impacts and radiation. Sports goggles are also recommended for use by tennis players and other racquet sports, like hockey and lacrosse. Cricket wicket keepers often wear helmets to protect their eyes from injury as a result of being hit by the cricket ball. Even serious fly fishermen should protect their eyes against fish-hooks that can penetrate their eyes.

5.7.1.3 Face Protection and Mouthguards

Fractured facial bones, and teeth (Fig. 5.12a), are common amongst cricketers, boxers, football and hockey players where the player is either hit by an accelerating ball, by a racquet or by an opponent or stress fractures from repetitive blunt force in boxing. Batting helmets and face guards are used to protect against such injury. The mouth, lips, teeth, gums, jaws, tongue and cheek are vulnerable to blows that can cause tears, fractures and even concussion depending on the impact [16]. Mouth protection (Fig. 5.12b) is a requirement for sports, like boxing, hockey, rugby and squash, where collision and trauma may be high. These guards should fit the mouth appropriately, be durable and adequately cleaned between wear for activities.

5.7.1.4 Pads, Guards and Straps

In any contact sport, like hockey or rugby, pads and guards should be worn to reduce injury to the neck, shoulders, chest, elbows, arms, wrists, hip, thighs, knees, shins and ankles. Guards range from hard plastic to soft padding, depending on the type of sport and expected injuries. Cricket players wear shin guards to protect the shins from rapid contact by the hard cricket ball, as well as

Fig. 5.13 Football shoes design for top-level football players

football players to protect against direct contact from opponents. Kneepads protect damage to cartilage and the knee joint, while shoulder pads can help to support the shoulder joint to reduce risk of sprains and fractures. Thigh pads are worn in cricket to prevent ruptures or severe bruising to the thigh muscles. Elbow pads are often worn in racquet sports too, such as tennis, but particularly in hockey and inline skating.

5.7.1.5 Protective Clothing and Footwear

Padded or reinforced clothes are uniquely designed for certain sports, as are the sports shoes. In field sports, such as football [17] or football, sports cleats with plastic spikes are worn for traction and to reduce the risk of injuries from falls. Running shoes, and ankle supports, are worn by athletes and are designed for optimised pronation and minimise overuse of the feet and ankle joint (Fig. 5.13). Cyclists wear cycling shoes that protect the feet from pain and allow for safe, fast pedaling.

References

1. Gleeson M (2002) Biochemical and Immunological markers of overtraining. J Sports Sci Med 1:31–41
2. Iliukhina VA, Tkachev VV, Fedorov BM, Reushkina GD, Sebekina TV (1989) Omega-potential measurement in studying the functional status of healthy subjects with normal and hypertensive types of reaction to graded physical exertion. Fiziol Cheloveka 15:60–65
3. Kääb MJ, Ito K, Rahn B, Clark JM, Nötzli HP (2000) Effect of mechanical load on articular cartilage collagen structure: a scanning electron-microscopic study. Cells Tissues Organs 167(2–3):106–120
4. Honda K (2000) The effects of high magnitude cyclic tensile load on cartilage matrix metabolism in cultured chondrocytes. Eur J Cell Biol 79(9):601–609
5. Fujisawa T et al (1999) Cyclic mechanical stress induces extracellular matrix degradation in cultured chondrocytes. J Biochem 125:966–975
6. Nigg BM (1985) Biomechanics, load analysis and sports injuries in the lower extremities. Sports Med 2(5):367–379
7. Hubscher M, Zech A, Pfeiper K et al (2010) Neuromuscular training for sports injury prevention: a systematic review. Med Sci Sports Exerc 42(3):413–421

8. Melnikov A, Singer RN (1998) Analysis of event-related potentials to filmed action situations. Res Q Exerc Sport 69:400–405

9. Yang JC, Lee CW (1988) Physique, physical characteristics, body composition analyses in various sport athletes – focusing 88 Olympics National Team athletes. Korean J PhysEduc 27(1):285–312

10. Matsudo VKR (1996) Prediction of Future Athletic Excellence. In Bar-Or, O (ed), The Child and Adolescent Athlete. Oxford: Blackwell Science. pp. 92–109

11. Ko B-G, Kim J-H (2005) Physical fitness profiles of elite ball game athletes. Int J Appl Sports Sci 17(1):71–87

12. Bangsbo J, Iaia FM, Krustrup P (2007) Metabolic response and fatigue in football. Int J Sports Physiol Perform 2(2):111–127

13. Witvrouw E, Mahieu N, Danneels L, McNair P (2004) Stretching and injury prevention: an obscure relationship. Sports Med 34(7):443–449

14. Kubo K, Kanehisa H, Fukunaga T (2002) Effect of stretching training on the viscoelastic properties of human tendon structures in vivo. J Appl Physiol 92(2):595–601

15. Ellis TH (1991) Sports protective equipment. Prim Care 18(4):889–921

16. Knapik JJ, Marshall SW, Jones BH et al (2007) Mouthguards in sport activities: history, physical properties and injury prevention effectiveness. Sports Med 37(2):117–144

17. Lees A, Kewley P (1993) The demands on the football boot. In: Reilly T, Clarys J, Stobbe A (eds) Science and football II. E & FN Spon, London, pp 335–340

Special Aspects of Prevention in Children and Adolescents

6

Antonio Maestro, Gorka Vázquez, Manuel Rodríguez, and Xavier Torrallardona

Key Points

1. Children and adolescents differ from adults with different injury patterns based on their anatomy and stage of development.
2. Targeted risk individual factors should be essential. Special control over individual intrinsic factors is a priority.
3. If workload is not considered, adolescent growth spurt could determine a high incidence of overuse injuries.
4. Orthopaedic surgeons should promptly identify the nature and extent of the injury, in order to decrease the risk of permanent disability.
5. Parents and coaches should also be involved in sports safety education, although primary-care physicians and paediatricians are more often in a position to provide anticipatory counselling.

6.1 Definition of "Child" Versus Definition of "Adolescent"

It is difficult to establish universal limits between childhood and adolescence, but in strictly semantic terms, childhood is the period between the end of infancy and adolescence, while adolescence begins with puberty or preadolescence (the term preadolescent refers to boys and girls who have not yet developed secondary sex characteristics) and ends with adulthood (Table 6.1), a time when both sexes reach reproductive maturity: semen ejaculation or spermarche in males and menarche in females [14, 30, 53].

Childhood is the stage from the child's second year of life to the beginning of adolescence and can be divided into two stages:

- The *pre-school stage*, from 2 to 6 years. In this stage, the child continues his rapid growth and development, the musculoskeletal system is still largely immature, and movement patterns are still quite basic due to an immature neuromuscular system.
- The *school stage*, from the ages of around 6 to 12 in girls and around 6 to 14 in boys. This

A. Maestro, MD, PhD (✉)
FREMAP, Avda. Juan Carlos I, 1, Gijón 33212, Spain
e-mail: doctorantoniomaestro@gmail.com

G. Vázquez, Pt, DO
Human Anatomy and Embryology Department,
University of Oviedo,
Julián Clavería, s/n, 33006 Oviedo, Spain
e-mail: dogorka@gmail.com

M. Rodríguez, MD • X. Torrallardona, MSc
Sport & Health Center,
Balneary Las Caldas, Oviedo 33174, Spain
e-mail: mrodriguez@lascaldas.com;
xavi.torrallardona@gmail.com

© ESSKA 2016
H.O. Mayr, S. Zaffagnini (eds.), *Prevention of Injuries and Overuse in Sports: Directory for Physicians, Physiotherapists, Sport Scientists and Coaches*, DOI 10.1007/978-3-662-47706-9_6

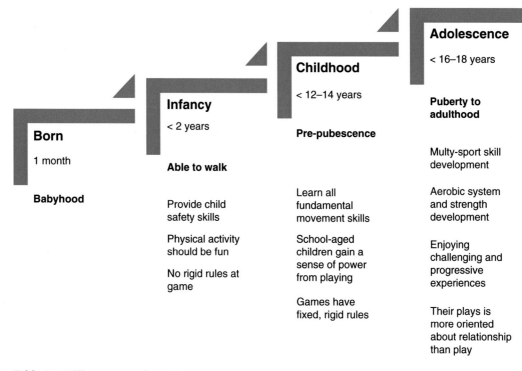

Born

1 month

Babyhood

Infancy

< 2 years

Able to walk

Provide child
safety skills

Physical activity
should be fun

No rigid rules at
game

Childhood

< 12–14 years

Pre-pubescence

Learn all
fundamental
movement skills

School-aged
children gain a
sense of power
from playing

Games have
fixed, rigid rules

Adolescence

< 16–18 years

**Puberty to
adulthood**

Multy-sport skill
development

Aerobic system
and strength
development

Enjoying
challenging and
progressive
experiences

Their plays is
more oriented
about relationship
than play

Table 6.1 Different stages of growth

stage shows relative stability in the progression of physical growth and maturity of the musculoskeletal and neuromuscular systems.

Both stages show minimal maturity of the central and peripheral nervous system; therefore, the neuromuscular and cardiorespiratory systems are not prepared to receive high volume (the total amount of activity performed by the subject during an exercise, session or work cycle), high-intensity workloads (the total stimulus applied on the body during physical activity can differ from external load associated with the volume and intensity of training) and internal load (the organic effect within the body as a consequence of the external load). As such, the recommended load of physical activity is low, such as games focused on developing neuromuscular coordination and light aerobic and anaerobic exercise.

Adolescence involves the transition from childhood to adulthood. It is time of enormous growth in height, so a great variation exists among children in size, body composition, rate of growth and physical maturation—in summary, significant changes on a physical, mental and social level [14, 53].

Adolescence coincides with the appearance of sexual development traits. It is characterised by the adolescent growth spurt, during which children gain physical, mental and emotional maturity in a very short time.

In boys, for example, peak velocities in lean mass, bone mineral and strength systematically follow peak linear growth (peak height velocity), in that order.

As for girls, menarche is coincident with peak bone mass following after peak height velocity and peak weight velocity, in that order. The common range in the onset of these events is high [3, 58].

Based on all of these factors, we could define childhood as the phase of bodily and psychological growth that precedes puberty or sexual development, while adolescence is the continuity of the process of adaptation and stability of maturation up until adulthood.

If we consider that child development is not only conditioned by physical growth, psychological aspects shall be taken into account when designing strategies and injury prevention programmes.

From the perspective of physical activity in childhood and adolescence, two main goals are to be achieved in both stages: to improve muscle strength and to develop fundamental motor skills [48], by doing a variety of exercises with progressive workloads that are consistent with individual needs, goals and abilities. As these exercises should precede intense training, the warm-up phase of physical activity in schools should be adapted accordingly as common practice.

6.2   Differences of the Child's Physis Compared to Adults

It is known that a child is NOT an adult, with the main differences in the musculoskeletal system being immature bone in terms of both of its composition (rich in organic matter and water and a relative mineral deficiency) and its architectural structure, elastic joints (in terms of both joint cartilage and capsuloligamentous elements) and a lesser likelihood of muscle hypertrophy, for which at this particular juncture the system is still fragile while being simultaneously influenced by hormonal, nutritional and mechanical factors which give it even greater structural variability [10, 14, 21, 27, 39, 53–58].

The growth plate is only present at children's bones. The child grows thanks to the physis or *growth cartilage* found in long bones and which appears at different points in time depending on the location [3, 14].

Being aware of huge variability on physis fusion appearance, we could consider that until the age of 10, the growth process occurs largely in parallel in both boys and girls. By contrast, girls undergo a growth spurt between the ages of 11 and 13, while this occurs in boys between 13 and 19, coinciding approximately with the hormonal changes of puberty and the physeal closure which is no longer active at 16–17 in girls and at 17–18 in boys [2, 3, 14, 53].

The structural development of bone originates in the growth plates of long bones, and this growth plate is avascular and aneural and consists of cells (chondrocytes) embedded in an abundant extracellular matrix. These cells are contained in the growth plate at different stages of differentiation, which are organised and arranged into several zones [22, 29, 58].

The growth plate (bone, periosteum and cartilage) is less resistant to stress than adult articular cartilage [22], and also, it is less resistant than adjacent bone to shear and tension forces. The physis may be 2–5 times weaker than the surrounding fibrous tissue [29]. Therefore, when disruptive forces are applied to an extremity, failure could occur through the physis. So physeal injuries may produce irreversible damage to the growing cells, leading to growth disturbance [15, 27].

There is no doubt that microtraumas caused by sport are repetitive by nature and that when they act on nonvascularised or poorly vascularised structures, such as elements of the joints, certain layers of growth cartilage, epiphyseal ossification centres or apophyses with tendinous, aponeurotic insertions become accumulative and thus acquire the ability to trigger the overload injury.

When practising sport, the growing osteoarticular system and, more specifically, the musculoskeletal system are subject to exogenous and endogenous mechanical stresses of pull/pressure, flexion and torsion, namely, self-trauma and microtrauma, which have repercussions on those same structures by triggering inflammatory processes that result in endochondral ossification and subsequent growth [57].

Repetitive altered or asymmetrical loading over growth plate cartilage could produce many skeletal deformities [15, 27]. Premature fusion may create a shortening of the limbs. If this shortening is in the upper limbs, the functional consequences tend to be minimal or insignificant. Load bearing could produce more significant effects on lower limbs, like leg-length inequalities, torsions and deformities resulting from malunion of fractures.

Several authors [8, 15, 27] have described that certain sports, like gymnasts, have altered relative growth of the radius and ulna which could be associated with repetitive growth plate injuries and/or altered loadings selectively retarding radial or ulnar growth, while in asymmetry sports like tennis [27], professional players had wider bones on their playing-side arms. They have also described that bone length differences between dominant and nondominant arms

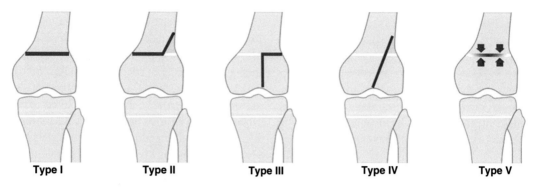

Fig. 6.1 Schematic drawing of epiphyseal injuries (Adapted from Salter and Harris Classification [46])

on the playing-side ulnar and 2nd metacarpal were, respectively, 3 % and 3.7 % longer than on the nondominant side, while control subjects presented no measurable differences.

As the joint, and in particular the growth cartilage, is the weak link in this assembly, it is thought that the risk of injury could be increased at this site during the growth spurt [2, 3, 34, 38] (Fig. 6.3). The muscle-tendon elongates in response to longitudinal growth in the long bones of the extremities, which could produce a temporary disparity between muscle-tendon and bone lengths. Therefore, it has been suggested that the growth spurt may also increase susceptibility to growth plate injury by causing an increase in muscle-tendon tightness about the joints and an accompanying loss of flexibility [35], and if excessive muscular stress is applied, a muscle-tendon imbalance is created which may predispose to injury [11, 13]. On the other hand, it has been questioned whether a reduction in flexibility happens during the adolescent growth spurt [21].

6.2.1 Acute Physeal Injury

Salter and Harris [46] designed a system (Fig. 6.1) that is generally used: Evolutive prognosis of each type of fracture depends on the blood supply, the presence of germinal cells and the fracture's displacement. For type I and II fractures, as opposed to what has been initially believed, these injuries are not as innocuous and can be associated with risk of growth impairment [38, 39, 46]. If germinal cells remain with the epiphysis, and circulation is unchanged, prognosis can be more favourable.

Type III injuries have a good prognosis if the fracture is not displaced, the blood supply is unchanged and the separated portion of the epiphysis is still undamaged. Surgery is in some cases necessary to restore the joint surface to normal. In type IV injuries, surgery is needed to restore the joint surface to normal and to align seamlessly the growth plate. Type IV injuries have a poor prognosis unless the growth plate is totally and accurately realigned.

Approximately 15 % of all fractures in children involve the physis. Injuries that would cause a sprain in an adult can be a potentially serious growth plate injury in a child [28, 35, 38, 39].

6.2.2 Chronic Physeal Injury

Several authors [8, 15, 27, 57] have described sports training as a main factor to precipitate pathological changes of the growth plate and, in extreme cases, cause growth disturbance. This chronic injury is very common in children and appears to happen through repetitive loading, like physical activity of high intensity or duration, which disturbs the mineralisation of the hypertrophied chondrocytes in the zone of provisional calcification (hypertrophic zone continues to widen because of constant growth in germinal and proliferative areas) and interferes with metaphyseal perfusion [38]. Stress fractures or osteochondroses (Osgood-Schlatter disease,

Table 6.2 Location and eponyms for osteochondroses

Center	Location	Eponym
Primary	Carpal scaphoid	Preiser disease
	Lunate	Kienböck disease
	Medial cuneiform	Buschke disease
	Patella	Köhler disease
	Talus	Mouchet disease
	Tarsal scaphoid	Köhler disease
	Vertebral body	Calvé disease (Legg-Calve-Perthes Disease)
Secondary	Vertebral epiphysis	Scheuermann disease and Scheuermann kyphosis
	Iliac crest	Buchman disease
	Symphysis pubis	Pierson disease
	Ischiopubic junction	Van Neck disease or phenomenon
	Ischial tuberosity	Valtancoli disease
	Calcaneal apophysis	Sever disease or phenomenon
	Accessory tarsal navicular or ostibiale externum	Haglund disease
	Second metatarsal	Freiberg disease (or Freiberg infarction)
	Fifth metatarsal base	Iselin disease
	Talus	Diaz disease
	Distal tibial epiphysis	Lewin disease
	Proximal tibial epiphysis	Blount disease
	Tuberosity of the tibia	Osgood-Schlatter disease
	Secondary patellar center	Sinding-Larsen-Johansson syndrome (Sinding-Larsen disease, jumper's knee)
	Lesser trochanter of the femur	Monde-Felix disease
	Greater trochanter of the femur	Mandl or Buchman disease
	Capital epiphysis of the femur	Legg-Calve-Perthes disease
	Phalanges	Thiemann syndrome
	Metacarpal heads	Mauclaire disease
	Proximal epiphysis of the radius	Schaefer disease
	Distal epiphysis of the ulna	Burns disease
	Medial humeral condyle	Froelich disease
	Lateral humeral condyle	Froelich disease
	Capitellum of the humerus	Panner disease (see Little League Elbow Syndrome)
	Humeral head	Hass disease
	Clavicle	Friedrich disease

Sever's disease) are the most common overuse injuries during childhood and adolescence (Table 6.2).

These alterations may cause asymmetric or irregular growth, or they may involve the entire physis and result in an overall slowdown of the rate of growth or even complete cessation of growth. In either case, premature closure of some or all of the physis may occur.

6.3   Functional Anatomical Aspects of Children and Adolescents

There are many anatomical differences between children and adolescents if we compare them with adults, which is why practising adult-level physical activity as a youth can result in injury and disease. The ongoing change to their anthropometric and organic components produces significant differences in the different stages of development between childhood and adulthood.

The most notable differences are at the osteoarticular level, and as this is still a developing system, it is physically vulnerable to loads, which may not seem excessive in other circumstances. The existence of other areas of growth, which are fragile under the mechanical stresses they receive, as well as the accelerated and often irregular growth of musculotendinous structures, can cause injuries.

At the organic level, there are other differences which also facilitate the occurrence of injuries and overloads during development, including thermoregulatory, cardiorespiratory, endocrine, nervous and adipose tissue (BMI) systems, and so on. Not to mention biomechanical and postural aspects typical of these ages and deriving from factors occurring prior to axis alignment.

It is vital to identify intrinsic factors of the child, namely, flexibility, elasticity, leg-length differences, axes of the limbs, nutritional state, systematic health problems, or poor prior rehabilitation, as these will be associated with extrinsic factors, namely, ground surface (natural grass, artificial turf or outdoors hard track) (Fig. 6.2), footwear, climate, trauma, loads, and poor technique, which are the main causes of injury.

The inverse relationship between the intensity of the microtrauma and the capacity of absorption on the surface of contact (i.e. hardness of the ground) must be taken into account [1, 26].

Probably, the risk is increased at adolescent growth spurt due to injuries attributable to such factors as decreased physeal strength and considerable augmentation of muscle-tendon tightness [13, 35]. The reduction of physeal strength is promoted by a lag of bone mineralisation behind bone growth during pubescent spurt, thus rendering the bone temporally more porous and more prone to injury during this stage [2, 57]. It has been observed that vigorous activity at early ages can cause a greater risk of bone deformities, although the veracity of this hypothesis is still being studied to determine whether intense activity during adolescence can abnormally alter the growth plate in the epiphyseal extension before and/or after closure of the physis [50].

In children, both tendon and ligament structures are strong and elastic, enabling them to withstand mechanical loads better than their areas of insertion, resulting in bone deterioration in the event of intense trauma or repeated microtrauma. For this reason, injuries to the growth cartilage and epiphysis are often observed alongside a low incidence of ligament injuries [35].

Moreover, dietary and nutrient recommendations for adolescents are needed in order to ensure correct nutrition to optimise bone growth and consolidation during this important life stage [58]. An adequate dietary intake of calcium, vitamin D, proteins and other minerals, as well as iron for girls, shall largely determine the correct osseous, muscular and organic growth in adolescents [10, 18, 33, 34].

Fig. 6.2 Extrinsic factor as ground floor (Artificial turf or outdoors hard track) should be considered as one of the most important when overuse injury is present

Body mass index (BMI; kg/m^2) is probably the most commonly used indicator for anthropometry assessments, as it is easy to use in measurements and shows a reasonable relation to body fatness in the general population. Heavier weight causes greater forces, which are absorbed through soft tissue and joints, thus perhaps related to an increased risk of injury. To this respect, several authors have reported an increased rate of injury among heavy players or players with high BMI [1, 10].

Therefore, it is of critical importance that the total caloric intake and quantities of macro- and micronutrients are adapted according to the needs of the child or adolescent.

Historically, observations of functional and morphological differences have indicated a less effective thermoregulation in children when exercising in a hot environment since they have a smaller *body surface* area than adults, with the problem occurring when exercising in extreme temperatures due to excess heat gain or heat loss in cold situations.

Heat shocks and heat exhaustion, particularly in hot climates, are more likely to occur in children than in adults, because children produce more heat relative to body mass, their sweating capacity is low, and they also tend not to drink enough compared to adults [11]. For this reason, to be able to establish a correct guideline of hydration during and after physical activity (drink 1.5 L of water for each kilogram of weight lost during activity) is important to reduce the occurrence or recurrence of injury. This practice should be followed in training as well as in competition

Strength is the amount of tension that can be produced by a muscle in a voluntary contraction. During pubescence, young boys and girls commence to increase muscle strength and demonstrate a high capacity and responsiveness to an increased training load, but it is important to note that in this stage it is easier to create an exercise overload that may induce overtraining; however, their bone structure is less stable and can be injured easier than in previous phases due to excessive loads or training volume [5].

It is important to underline that resistance training exercise, involved at multifaceted pro-gramme of physical activities, should commence at prepubertal ages and should be maintained throughout the pubertal development, to obtain a well-developed bone structure activity before pubertal growth spurt. This resistance programme training stimulates both bone and skeletal muscle hypertrophy to a greater degree than observed with normal growth in nonphysically active or sedentary children [48, 54, 56].

6.4   The Importance of the Coach, the Family and the Social Environment

Regardless of the chosen sport, the main concern of parents, trainers and health professionals has to be the health and wellbeing of the child. It is essential to plan childhood sporting activity carefully, as a poorly guided or practised sport can have harmful physical and psychological consequences, in some cases irreversible.

All people in the child's environment (family, coach, physiotherapists, doctors, etc.) and who have a more or less significant involvement in the planning of the child's physical activity must at least be aware of all aspects which could, at any given time, expose the child or adolescent to injury or any other negative effect deriving from the practice of sport (risk factors, injury mechanisms, etc.). Ideally, they should also be familiar with the positive effects of physical activity on the body of a child or adolescent, both physiologically and psychologically.

Coaches need to influence athletes through guidance and teaching, providing a safe learning environment. The experience of the coach is key to preventing injuries, and intuition suggests that poor coaching may be a predictor for injury.

As part of their responsibilities during practice, coaches interact directly with athletes to improve their sporting ability and take measures to guarantee their safety [47]. The responsibilities of the coach outside practice include numerous tasks which supplement everyday training, such as relationships with parents, working with other coaches, regular meetings with the facility or club's medical department, etc.

A good relationship between the doctor, coach, child and parents is essential for the child to enjoy a pleasant and successful experience in sport. Parents need to be coached on their roles and responsibilities within the sport. This is only achieved through complete communication, and it is vital that parents are aware of the appropriate types of sport for each stage of their child's development, as they know their child best. They must therefore advise them when choosing a sport and on its implications.

There is a chronological sequence of physical activities in children and adolescents approved by the scientific community for beginning coached physical activity, which may be established from.

Pre-school Stages

Ages 2–6, no impact exercise, like swimming or cycling (except in children with Scheuermann's disease), aimed at strengthening the muscle and vertebral structure and as a respiratory exercise, but never as a competitive sport. It is important to synchronise and strengthen the musculature of the limbs.

School Stages

Ages 7–9, low-impact exercise. Ballet, athletics (running, racewalking), synchronised swimming, rhythmic gymnastics and hiking are the sports generally advised at this stage.

Ages 10–12, medium-impact sports. Football, handball and basketball, for instance, can be played with a certain level of precaution. Asymmetric sports like tennis, canoeing, hockey or throwing should be considered separately. At this age, the child may begin competitive sports at a reasonable pace, but under close control by specialists in sports medicine and qualified coaches.

Sports which involve the use of a predominant single hand such as racquet sports, handball or fencing should be carefully monitored as they place stress on the muscles of only one side of the body, potentially resulting in a morphological imbalance. This can be avoided with the appropriate compensatory exercises.

At present, one of the most recommended and complete ways to reduce injuries is the pre-participation physical examination, as well as the first step in injury prevention. An accurate medical history will determine any pre-existing medical problems and should serve as the cornerstone of the examination. It is of vital importance that parents get involved in this process because only they can contribute to complete the medical history [42].

Any child experiencing an accelerated phase of growth must reduce workloads proportionately; otherwise, we take the risk of injuring the weakest areas of the underdeveloped skeleton, growth cartilage, epiphysis, articular cartilage and insertional apophysis, which are the most commonly damaged areas and must be given special attention if symptoms should occur.

In many cases, optimistic expectations surrounding an athlete's potential create a "support campaign" environment which in most cases leads to errors: imitation of work systems which are inappropriate for the child's chronological age and maturational status, nonprogressive increase in the density or frequency of sessions, nutritional changes that are not suitable for a phase of weight growth or a combination of two or more simultaneous competitions. These among other factors can increase physical and psychological stress levels, which can result in injury or even quitting the sport. For this reason, both the coach and the family must always assess the child's development objectively and consult his doctor in regard to speeding up or increasing his workloads.

Muscular imbalances and segmental misalignments of the skeleton, as well as changes in eating habits, are processes which, with adequate training, parents and coaches can identify and put the child in contact with the right professionals (orthopaedic surgeon, physiotherapist, nutritionist, etc.).

Parents' help is vital in implementing childhood injury prevention programmes, being heavily responsible for promoting healthy lifestyle habits (including exercise, nutrition, recovery, personal hygiene, etc.). Habits established in childhood often continue into adulthood. Besides the rapport between parents, coaches and physicians should be ongoing.

In case of injury, the medical team must inform the coach and parents of the current injury management strategy and recommended steps to prevent future injury.

Therefore, sports recommendations for children and adolescents must be controlled both by a medical team to monitor intrinsic factors and by the coach who establishes a routine that is appropriate for the child's stage of development, reducing training load in stages of rapid growth in order to avoid functional overload and frequently structuring any information provided by parents in regard to their assimilation by the adolescent [10].

6.5   The Child's Protection Against Overload

Nowadays, children's sports seasons last longer and are more intense, and many young athletes take on numerous sports at once, increasing their risk of injury.

Overuse injuries comprise a broad spectrum of injuries within sports medicine and are defined as chronic injuries related to constant levels of physiological stress without enough time to recover. Therefore, they are a microtraumatic type of injury to bone or soft tissue that is subject to repetitive stress without a long healing time. They can be perceived as the result of the difference between the volume of the stress or force applied to the body and the ability of the body to dissipate this stress or force. Fifty percent of all injuries seen in paediatric sports medicine practices could be attributed to overuse [6].

The mechanism of action of overuse injuries can result from repetitive microtrauma imposed on otherwise healthy tissue or the repeated application of lesser magnitudes of force to pathologic tissue [40].

All overuse injuries are not intrinsically innocuous. Doctors must be familiar with specific high-risk injuries, including femoral neck stress fractures, tarsal navicular fractures, fractures of the anterior tibial cortex and of the physis, femoropatellar pain syndrome, etc. [16].

When an overuse injury is diagnosed, it is essential to identify the underlying cause. The athlete, parents and coaches must participate in reviewing all the risks and factors to develop a strategy for preventing recurring injuries.

Overuse injuries are common in children and adolescents who practice sport regularly. In young athletes, these injuries are the result of a complex interaction of multiple factors, including growth-related factors which are unique to this population, although to self-limit such injuries, recovery times may be extended, often more than for acute injuries.

The prevention of overload injuries is probably the most important aspect of any sports programme. Those entrusted with monitoring the physical/sporting activity and the health of children and youths must consider this potential risk when creating such programmes.

Traditionally, risk factors for overuse injuries are classified as either intrinsic or extrinsic risk factors, being both susceptible to increase the likelihood of this type of injuries.

Intrinsic risk factors are classified as implicit or unique to the individual that may increase the likelihood of sustaining an injury. Maturational status, body mass index (BMI), gender, anatomic variations, as well as biomechanical movement patterns are all examples of intrinsic risk factors.

Extrinsic risk factors are those factors that, when applied to the athlete, could increase the risk of injury. These factors may include training methods, equipment and environment and may affect the magnitude, stress, or force applied to the body.

Early participation in specific sports has grown during the last four decades, reflecting a general trend towards specialisation and competition to childhood and adolescents. This intensity at early stages produces injuries due to excess activity without significant control, such as patellofemoral knee pain, stress fractures and osteochondritis dissecans (OCD) [12, 23, 25, 32].

Meanwhile, the increased competitive element exposes children to more serious injuries such as contusions and injuries to the bone physis and anterior cruciate ligament [4, 43] for which it must be acknowledged that each child has individual risk factors and each sport carries its own risks [34].

Special attention must be paid to musculotendinous imbalance, since the stress of growth is accompanied by a reduction in flexibility and an increase in risk factors due to overuse which can cause insertional apophysitis, for which stretching, flexibility and progressive strengthening work is required to stabilise the musculotendinous balance of large joints.

Another aspect that promotes overload injuries is weakness in muscle groups involved in complex movements (jumping, running, etc.). This causes other muscle, tendon, ligament and joint structures to assume an extra workload for which they are not prepared nor designed, exceeding over time their recovery capacity and resulting in injury. This imbalance due to the alteration of movement patterns is more frequently seen in the lower limbs.

The most common imbalances are objectified to test the strength of agonist and antagonist muscle groups in the same motor pattern. As seen on the table, traditionally the interest on muscle imbalances was mainly focused on knee flexors and extensors, but it was only during the last decade that emphasis has shifted towards assessing hip strength, hip muscular activation and running gait mechanics to assess the relative impact these variables may have on the development of a running-related injury. A lack of strength in the hip abductor, hip extensor and hip external rotator musculature is theorised to place the femur into excessive amounts of adduction and internal rotation resulting in alterations in joint coupling and mechanics to the knee, shank and foot-ankle complex distally. The aforementioned lack of hip strength has been associated with different and common injuries on children, including patellofemoral syndrome, iliotibial band syndrome and tibial stress fracture [41, 54].

It is also common during growth to observe misalignments of body segments [pes cavus (high arch) or pes planus (flatfoot), genu varum, valgus, hyperextension, scoliosis, hyperlordosis, kyphosis, etc.]. These may be due to a hereditary component and/or a consequence of an imbalance in the musculoskeletal system. The most common mechanisms are asymmetric or intense pressures on growth plates, an imbalance in agonist-antagonist muscle forces, reduced flexibility and/or increased muscle tone.

These anatomical alignment defects contribute to the appearance of a stress disorder, with

tendons or ligaments and even bones having to withstand stresses far greater than those of a normal skeleton. The skeletal muscle system is designed to withstand loads based on a given physiological posture; when we stray away from this balance, the forces applied fall on structures that are not specifically prepared giving way to premature overloading or overuse. An increased incidence of knee injuries in runners with a Q angle greater than 20° has been observed, while runners with more than a 4° Q angle difference were more likely to injure their shin [44].

It has been postulated that flatfoot type results in an increase in pronation excursion and excessive strain to medial soft tissue structures, whereas a high-arch type results in a stiffer foot that is less well equipped to dampen ground reaction force (GRF) at the foot and ankle leading to excessive bone stresses and lateral column injuries [40, 45].

In the lower limb, diseases such as Osgood-Schlatter, Sinding-Larsen-Johansson and Sever's could be associated with incorrect practice of sports due to both extrinsic (surfaces, footwear, inadequate nutrition, routine and/or functional overload errors) and intrinsic (rapid growth, muscular imbalance, postural alterations, flexibility deficit, ligamentous hyperlaxity, dysmetria of the lower limbs, metabolic changes, muscular fatigue, emotional changes) factors. In order to help dampen these external forces and reduce the likelihood of traumatic tissue failure, many injury prevention programmes attempt to develop the athlete's neuromuscular control mechanisms, but there is significantly less evidence regarding the most efficacious programme to decrease the incidence of these overuse injuries [6, 16, 35].

In order to avoid serious damage to the musculoskeletal system of young athletes, it is very important to gain in-depth experience and knowledge of the different aspects of training, including intensity (i.e. the ease or difficulty of a training session), frequency (i.e. number of training sessions per week) and recovery.

Simply put, we could say that before they first step into the competitive world, it is important that children carry out at least one hour of game-based exercise in order to improve their physical conditioning for sports, given that preseason conditioning works well to reduce early season injuries [24].

It has been suggested [26] that when aiming at reducing the incidence of injuries and enhancing athletic performance of children beginning any sports activity, it is recommended to include during the warm-up and warm-down phase a short-volume but well-designed exercise programme that comprises neuromuscular and intermittent training.

Special consideration must be given at these ages to strength training, as this essentially involves concentric (the insertions of the muscle come closer) and isometric (the muscles contract but the ends do not move) contraction, although the eccentric and plyometric training technique should be taught from an early age, since both provide significant benefits in injury rehabilitation exercises and must be known by the adolescent [19, 34, 40].

The assimilation of physical activity is a process which requires unavoidable phases of rest and relaxation. Increasing the density of weekly sessions and/or reducing recovery times between exercises must be avoided until recommended by qualified experts. In general, increasing the duration or intensity of training by more than 10 % per week should be absolutely contraindicated.

In endurance sports, the "10 % rule", consisting in increasing activity by 10 % each week [48] must be respected.

As previously mentioned, growth spurt phases in adolescence present a serious risk of damage to bone and growth plates; hence, physical activity needs to be monitored in order to reduce risk of long-term skeletal damage.

6.6 The Child's Protection Against Injury

The public health impact of injury in youth sport is great, and although physical activity is associated with a better quality of life and an overall reduction in mortality and morbidity, there remains an associated activity-related risk of injury and reinjury. Injury prevention is becoming a greater public health priority considering the long-term health impact and societal burden related to decreased levels of participation in physical activity with the known socioeconomic implication of degenerative processes. Research is of significant importance in promoting safe exercise participation by identifying risk factors for injury and reinjury [33] and in trying to avoid or reduce incidence of early osteoarthritis following many injuries in youth sport [18].

Equally important as prevention programmes are orthopaedic surgeons, who are very often involved in the diagnosis and treatment of these injuries, but less frequently in their prevention. Coaches, teachers and parents consider their physicians to be a valuable source of safety education; therefore, orthopaedic surgeons have a unique opportunity to provide injury prevention advice [42].

Therefore, training programmes must be planned according to the degree of physical and psychological maturity of the athletes in order to adjust the physical demands to their bodily changes during that stage [49].

Again, it shall be noted that there is an increased risk of injuries at peak height velocity growth, and it was found that boys with the highest incidence of injury were tall (>165 cm). Also, weak muscles are a risk factor in this respect. Authors have found that injured athletes had a weak grip (<25 kg), suggesting that skeletally mature, but muscularly weak, boys may have more chances of injury while playing football with peers of the same chronological age [1].

Despite of high incidence of this type of sports injuries, the vast majority are minor and do not necessitate medical attention. Very serious sports accidents in youth, such as brain or spinal cord damage, lesions of the heart or submersions resulting in invalidity or death, are exceptional [55].

As mentioned before, it is important to have a good knowledge of the developing musculoskeletal system to understand children's injuries. Tendons and ligaments are relatively stronger than the epiphyseal plate and considerably more elastic. Therefore, in severe trauma, the epiphyseal plate, as it is weaker than the ligaments, gives way. Consequently, growth plate damage is more common than ligamentous injury [28, 31, 35, 51] and high-resistance training may predispose children to an increased risk of injury if not properly supervised [19].

Many women who experience delayed puberty can end up with open epiphyseal plates, which are prone to injury. These women have a greater risk of acute, long-term skeletal injuries such as stress fracture and osteoporosis [17, 20, 36, 37].

As result of strength and flexibility deficits in the growing athlete, muscular imbalances, segment misalignments and reduced flexibility can often cause apophyseal injuries. We know that at growth spurt (Fig. 6.3), there is a decrease in flexibility due to relative bone lengthening. This predisposes to injury when appropriate stretching exercises prior to commencing sport are not carried out.

Several papers have shown that the distribution of the injury area differed significantly by age, with younger patients more likely to be seen for upper extremity injuries and older patients more likely to be seen for head, chest, hip/pelvis and spine injuries [9, 52].

The lower extremity was the most frequently injured body part. Approximately half of the younger patients (5–12 years old) were treated for overuse injuries and half for traumatic injuries; injuries were primarily bony in nature.

A significantly larger proportion of injuries in older children (13–17 years) were to the soft tissues when compared with younger athletes. The proportion of younger athletes treated for osteochondritis dissecans, fractures (including physeal fractures) and apophysitis was higher when compared with their older counterparts.

In lower limbs, bone overuse, representing excessive apophyseal load (e.g. Osgood-Schlatter disease, Sever's disease, apophysitis), and osteochondritis dissecans (OCD) were predominant (Fig. 6.4) [22, 23, 25, 49].

As for upper extremity injuries, the patients treated were disproportionately male [52].

Upper extremity injuries were largely traumatic and bony, including fractures and OCD, particularly for younger patients.

The predominant upper extremity injury among older patients (13–17 years) was also a fracture, followed by joint instability and

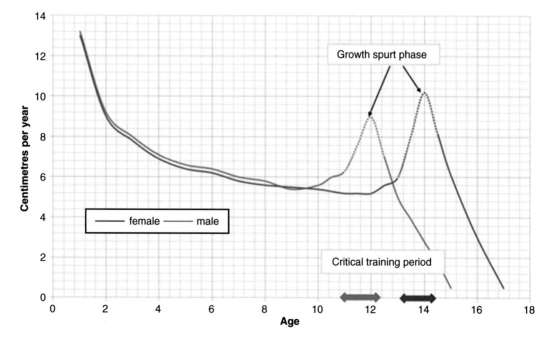

Fig. 6.3 Peak velocity height curve for girls and boys showing the growth spurt phase and critical training or exercise phases

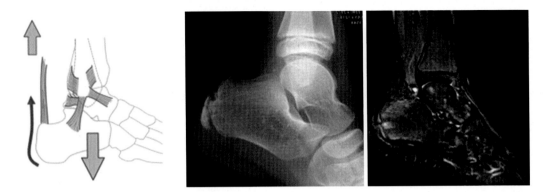

Fig. 6.4 Sever's disease in child football player

OCD. One of the most usual hip and pelvis injuries were labral tears, apophysitis and avulsion fractures.

A previous study [48] found that 70 % of the injuries involved the lower extremities, and regarding trunk injuries, 75.8 % of the younger patients that were treated for spine injuries were female (compared with 53 % in the full sample of younger patients). Most spine injuries were overuse and bony injuries. Spondylolysis accounted for the majority of the spine injuries.

There was no relevant difference observed in injuries to the spine when comparing the children (5–12-year age group) to the adolescent (13–17-year age group). Young female athletes with spine injuries were gymnasts, dancers or figure skaters, and male athletes were football, ice hockey and football players. Most of younger female patients (5–12 years) treated for lower extremity injuries were treated for overuse injuries that were bony in nature [52].

It is recognised that children and adolescents sustain more injuries from free play than from organised sports; therefore, knowledge of a game's rules and their observance play a vital role in preventing injuries.

It is of vital importance to use homologated safety sports equipment in order to reduce impacts or prevent the consequences of trauma in the field. Methods for decreasing football injuries include the use of shin guards, properly secured and padded goalposts, nonabsorbent balls for wet playing fields and proper cleat selection. Bicycle-related head trauma is an important cause of disability and death. Helmets can reduce the severity and incidence of head injuries, so the use of helmets that meet existing standards for bicycle helmets has been recommended for in-line skaters, as such helmets have been proved to be strongly protective against head injuries in physical environments very similar to those of skaters [42]. Wrist guards have been proved effective in the protection of in-line skaters in both case-control [12] and biomechanical studies.

A basic tool for prevention that should be considered is the improvement of rehabilitation programmes and their personalisation for each athlete and for each injury, as well as knowledge of the athlete's sensations prior to the injury. Additionally, information on the prevention and treatment of injuries for parents, coaches, trainers and therapists can reduce the number and duration of injuries [7].

References

1. Backous DD, Friedl KE, Smith NJ et al (1988) Soccer injuries and their relation to physical maturity. Am J Dis Child 142(8):839–842
2. Bailey DA, Wedge JH, McCulloch RG et al (1989) Epidemiology of fractures of the distal end of the radius in children as associated with growth. J Bone Joint Surg Am 71(8):1225–1231
3. Bailey DA, Baxter-Jones ADG, Mirwald RL et al (2003) Bone growth and exercise studies: the complications of maturation. J Musculoskel Neuron Interact 3(4):335–337
4. Bakhos LL, Lockhart GR, Myers RE et al (2010) Emergency department visits for concussion in young child athletes. Pediatrics 126(3):550–556

5. Bird EP, Tarpenning KM, Marino FE (2005) Designing resistance training programmes to enhance muscular fitness. Sports Med 35:841–851

6. Brenner JS, American Academy of Pediatrics Council on Sports Medicine and Fitness (2007) Overuse injuries, overtraining, and burnout in child and adolescent athletes. Pediatrics 119:1242–1245

7. Brophy RH, Backus S, Kraszewski AP et al (2010) Differences between sexes in lower extremity alignment and muscle activation during soccer kick. J Bone Joint Surg Am 92(11):2050–2058

8. Caine D, Caine C, Maffulli N (2006) Incidence and distribution of pediatric sport-related injuries. Clin J Sport Med 16:500–513

9. Caine D, Howe W, Ross W et al (1997) Does repetitive physical loading inhibit radial growth in female gymnasts? Clin J Sport Med 7(4):302–308

10. Caine D, Maffulli N, Caine C (2008) Epidemiology of injury in child and adolescent sports: injury rates, risk factors, and prevention. Clin Sports Med 27(1):19–50

11. Castiglia PT (1995) Sports injuries in children. J Pediatr Health Care 9:32–33

12. Cleland V, Timperio A, Salmon J et al (2011) A longitudinal study of the family physical activity environment and physical activity among youth. Am J Health Promot 25(3):159–167

13. Connolly SA, Connolly LP, Jaramillo D (2001) Imaging of sports injuries in children and adolescents. Radiol Clin North Am 39:773–790

14. Crasselt W (1988) Somatic development in children (aged 7 to 18 years). In: Dirix A, Knuttgen HG, Tittel K (eds) The Olympic book of sports medicine. Blackwell, Oxford, pp 286–299

15. De Smet L, Claessens A, Lefevre J et al (1994) Gymnast wrist: an epidemiologic survey of ulnar variance and stress changes of the radial physis in elite female gymnasts. Am J Sports Med 22(6):846–850

16. DiFiori JP, Benjamin HJ et al (2014) Overuse injuries and burnout in youth sports: a position statement from the American Medical Society of Sports Medicine. Clin J Sport Med 24:3–20

17. Dusek T (2001) Influence of high intensity training on menstrual cycle disorders in athletes. Croat Med J 42:79–82

18. Emery CA (2010) Injury prevention in paediatric sport-related injuries: a scientific approach. Br J Sports Med 44:64–69

19. Faigenbaum AD, Myer GD (2010) Resistance training among young athletes: safety, efficacy and injury prevention effects. Br J Sports Med 44:56–63

20. Feingold D, Hame SL (2006) Female athlete triad and stress fractures. Orthop Clin North Am 37:575–583

21. Feldman D, Shrier I, Rossignol M (1999) Adolescent growth is not associated with changes in flexibility. Clin J Sport Med 9:24–29

22. Flachsmann R, Broom ND, Hardy AE et al (2000) Why is the adolescent joint particularly susceptible to osteochondral shear fracture? Clin Orthop Rel Res 381:212–221

23. Flynn JM, Kocher MS, Ganley TJ (2004) Osteochondritis dissecans of the knee. J Pediatr Orthop 24(4):434–443

24. Heidt RS, Sweeterman LM, Carlonas RL et al (2000) Avoidance of soccer injuries with preseason conditioning. Am J Sports Med 28(5):659–662

25. Kocher MS, Tucker R, Ganley TJ et al (2006) Management of osteochondritis dissecans of the knee: current concepts review. Am J Sports Med 34(7):1181–1191

26. Koutures CG, Gregory AJM, McCambridge TM et al (2010) Injuries in youth soccer. Pediatrics 125(2):410–414

27. Krahl H, Michaelis U, Pieper HG et al (1994) Stimulation of bone growth through sports. A radiologic investigation of the upper extremities in professional tennis players. Am J Sports Med 22(6):751–757

28. Krueger-Franke M, Siebert CH, Pfoerringer W (1992) Sports-related epiphyseal injuries of the lower extremity. An epidemiologic study. J Sports Med Phys Fit 32:106–111

29. Larson RL, McMahon RO (1966) The epiphyses and the childhood athlete. JAMA 7:607–612

30. Lloyd RS, Faigenbaum AD, Stone MH et al (2014) Position statement on youth resistance training: the 2014 International Consensus. Br J Sports Med 48:498–505

31. Maffulli N, Baxter-Jones ADG (1995) Common skeletal injuries in young athletes. Sports Med 19(2):13–49

32. Maffulli N, Caine D (2005) The epidemiology of children's team sports injuries. Med Sport Sci 49:1–8

33. McBain K, Shier I, Shultz R et al (2012) Prevention of sport injury II: a systematic review of clinical science research. Br J Sports Med 46:174–179

34. Micheli LJ, Glassman R, Klein M (2000) The prevention of sports injuries in children. Clin Sports Med 19(4):821–834

35. Micheli LJ (1983) Overuse injuries in children sports: the growth factor. Orthop Clin North Am 14(2):337–360

36. Nattiv A, Loucks AB, Manore MM et al (2007) American College of Sports Medicine position stand. The female athlete triad. Med Sci Sports Exerc 39:1867–1882

37. Niemeyer P, Weinberg A, Schmitt H et al (2006) Stress fractures in adolescent competitive athletes with open physis. Knee Surg Sports Traumatol Arthrosc 14:771–777

38. Ogden JA (2000) Skeletal injury in the child, 3rd edn. Springer, New York

39. Pappas AM (1983) Epiphyseal injuries in sports. Phys Sportsmed 11:140–148

40. Paterno MV, Taylor-Haas JA, Myer GD et al (2013) Prevention of overuse sports injuries in the young athlete. Orthop Clin N Am 44:553–564

41. Powers CM (2003) The influence of altered lower-extremity kinematics on patellofemoral joint dysfunction: a theoretical perspective. J Orthop Sports Phys Ther 33(11):639–646

42. Purvis JM, Burke RG (2001) Recreational injuries in children: incidence and prevention. J Am Acad Orthop Surg 9(6):365–374

43. Radelet MA, Lephart SM, Rubinstein EN et al (2002) Survey of the injury rate for children in community sports. Pediatrics 110:1–11

44. Rauh MJ, Koepsell TD, Rivara FP (2007) Quadriceps angle and risk of injury among high school cross-country runners. J Orthop Sports Phys Ther 37(12):725–733

45. Razeghi M, Batt ME (2000) Biomechanical analysis of the effect of orthotic shoe inserts: a review of the literature. Sports Med 29(6):425–438

46. Salter RB, Harris WR (1963) Epiphyseal-plate injuries. J Bone Joint Surg Am 45(3):587–622

47. Schembri G (2001) Roles and responsibilities of the coach. In: Pyke FS (ed) Better coaching. Advance coach's manual, 2nd edn. Human Kinetics, Champaign, pp 3–13

48. Sewall BS, Micheli LJ (1986) Strength training for children. J Pediatr Orthop 6:143–146

49. Sharma P, Luscombe KL, Mafulli N (2003) Sport injuries in children. Trauma 5(4):245–259

50. Siebenrock KA, Ferner F, Noble PC et al (2011) The cam-type deformity of the proximal femur arises in childhood in response to vigorous sporting activity. Clin Orthop Relat Res 469:3229–3240

51. Stanish WD (1995) Lower leg foot and ankle injuries in young athletes. Clin Sports Med 14: 651–658

52. Stracciolini A, Casciano R, Friedman HL et al (2014) Pediatric sports injuries: a comparison of males versus females. Am J Sports Med 42:965–972

53. Tanner JM (1962) Growth at adolescence, 2nd edn. Blackwell, Oxford

54. Thorborg K, Couppé C, Petersen J et al (2011) Eccentric hip adduction and abduction strength in elite soccer players and matched controls: a cross-sectional study. Br J Sports Med 45(1):10–13

55. Tursz A, Crost M (1986) Sports-related injuries in children. A study of their characteristics, frequency, and severity, with comparison to other types of accidental injuries. Am J Sports Med 14(4):294–299

56. Vicente-Rodriguez G (2006) How does exercise affect bone development during growth? Sports Med 36:561–569

57. Wattenbarger JM, Gruber HE, Phieffer LS (2002) Physeal fractures, part I: histologic features of bone, cartilage, and bar formation in a small animal model. J Pediatr Orthop 22(6):703–709

58. Whiting SJ, Vatanparast H, Baxter-Jones A et al (2004) Factors that affect bone mineral accrual in the adolescent growth spurt. J Nutr 134:696S–700S

General Training Aspects in Consideration of Prevention in Sports

7

Karlheinz Waibel, Henrique Jones,
Christoph Schabbehard, and Bernd Thurner

Key Points

1. Training principles are the basic guideline for efficient, sustainable, and predictable adaptations.
2. A proper understanding of the needs and requirements of a specific sports activity in combination with a complex assessment of the individual performance abilities is a major component to establish effective prevention strategies.
3. Intra- and intermuscular coordination is determined by the interaction of the sensor systems and motor units.
4. Most physiological processes are dependent on the temperature of the surrounding milieu.
5. Anabolic and catabolic processes need to be balanced to prevent the organism from overuse or overstress symptoms.
6. To be aware of the hazards and the possible influences for a specific activity is the first step to initiate prevention strategies.

K. Waibel (✉)
German Ski Federation,
Hubertusstrasse 1, 82152 Planegg, Germany
e-mail: karlheinz.waibel@deutscherskiverband.de

H. Jones
Montijo Orthopaedic Surgery and Sports Clinic,
Rua Miguel Pais, 45,
2870-356 Montijo, Portugal
e-mail: ortojones@gmail.com

C. Schabbehard
Dipl. Sportlehrer,
Ringweg 2, 87763 Lautrach, Germany
e-mail: schabbehard@web.de

B. Thurner
Dipl. Sportlehrer,
Prevention Programs,
Therapy- and Training Center Friedberg,
Thomas-Dölle-Straße. 16,
Friedberg, Germany
e-mail: bthurner@therapiezentrum-friedberg.de

7.1 Training Principles

The focus of all efforts in elite sports and beyond is to optimize athletic performance, and therefore the athlete needs to be optimally trained. The collateral effect of a structured and purposeful process to achieve that goal is a high grade of prevention. For that it is worth to have a closer look to what are the basic requirements that make strength and conditioning activities be considered as training. Zintl [14] defined training as a methodical process with the goal of inducing the optimization, stabilization, or reduction of the complex physical and mental performance level of an individual. The essence of this definition is that training needs an aim and a plan how to achieve it.

© ESSKA 2016

H.O. Mayr, S. Zaffagnini (eds.), *Prevention of Injuries and Overuse in Sports: Directory for Physicians, Physiotherapists, Sport Scientists and Coaches*, DOI 10.1007/978-3-662-47706-9_7

The challenge for everyone is to determine the appropriate type of training not only in terms of exercise used, but most of all, to identify and understand the biological consequences of various training stimuli. The art of designing a training process means to understand how to apply the correct modality of exercise, the suitable balance of volume and intensity for each individual, and the correct timing of different interventions [10].

When designing training programs, it is important to have a clear understanding of the principles of training. These basic principles are described to help the athlete and coach set realistic goals and develop training programs that will provide the greatest opportunity to achieve performance and prevention gains [8]. Depending on the influence of the adaptation process, responding a training stimulus Zintl [14] distinguished three different categories.

7.1.1 Principles to Initiate the Adaptation

7.1.1.1 Overload Principle

Adaptation is a result of a stimulus that caused the homeostasis getting out of balance. The basis of the overload principle is that the intensity of the stimulus initiated by the executed exercise must be strong enough to perturb the biological balance of anabolic and catabolic processes of the muscle or other physiological components.

Following this idea, there are three types of training stimuli:

- Below threshold
- Optimal overload
- Excessive overload

From a preventive point of view, the third level of intensity can be problematic. An enduring overstress of the before-mentioned balance of biological processes may cause a damage of the specific physiological component. For a coach or an athlete, it is of great importance to have a proper knowledge about the individual performance level and the specific effect of the designed training program.

7.1.1.2 Progression Principle

Based on the ideas of the overload principle, the stimulating impact training on a specific level of intensity creates will attenuate over a certain period of time. To make sure that the training will further stimulate the targeted structures, the level of intensity needs to be permanently adjusted. This process of applying progressive overload occurs continually throughout a training program.

The challenge for both coach and/or athlete is to assess the individual performance level and adapt the training stimulus to reach the optimal grade of overload.

7.1.2 Principles Ensuring a Sustainable Adaptation

7.1.2.1 Smart Variation of Exercise and Rest Principle

This rule considers the fact that every overload stimulus needs a certain period of time of recovery before the next training exercise can be executed under almost equal or optimal conditions. Exercise and rest need to be considered as a unity when designing a training program. The background of this principle is that every biological system tries to reach and maintain the status of homeostasis. A stimulus set by executing a certain exercise makes this system lose balance and almost simultaneously the organism starts recovery processes to reattain this biological balance. As the organism strives to maintain the homeostasis, the recovery of the affected component will exceed the original level to avoid a status of imbalance induced trough an impact of similar character as the previous one (super compensation) [14]. The severity of the impact a specific training stimulus creates on the targeted physiological component is dependent on variables like motoric skill, intensity, or volume and on the individual preconditions of the athlete. Each specific impact requires a certain period of rest that needs to be considered when designing and

Table 7.1 Time of recovery in consideration of motoric action

Type of motoric action	90 % Recovery	100 % Recovery
Strength		
Max. strength	18–24 h	48–72 h
Explosive strength	18–24 h	48–84 h
Strength endurance	12–18 h	48 h
Endurance		
Aerobic	6–8 h	12–24 h
Anaerobic (lactate)	6–10 h	24–36 h
Speed/agility (PCr – metabolism + neuromuscular activity)	Approx. 10 h (max. intensity)	36–48 h (max. intensity)

planning a training program. Table 7.1 shows the recovery times for different training activities distinguished in full recovery and incomplete recovery. The recovery times displayed are determined for trained or active people and may double up for beginners. It is not generally necessary to wait for full recovery before inducing the next training stimulus. Experienced athletes commonly strive to create a deeper impact on the homeostasis of the targeted physiological component by cutting recovery time at a level of 90 % of complete reconditioning.

7.1.2.2 Continuity and Regularity Principle

To ensure that the investment in training and exercise finally pays out a continuous and regular activity is required. Progressive increase of the targeted performance is guaranteed as long as there are continuous specific training stimuli set. In case of an interruption of this continuity due to injury, lack of motivation, illness, irregularity, or excessive periods of rest, the performance gains will diminish. An important aspect concerns the timeline of those effects. The more rapid the performance gains were achieved due to hard work, the more rapid these adaptations diminish and even vanish if the proper stimulus lacks.

Effective prevention programs ought to be thoroughly implemented in the regular training routine of athletes in order to guarantee the sustainability of its effects.

7.1.3 Principles for Directing a Specific Adaptation

7.1.3.1 Individuality Principle

The principle of individuality in training refers to the fact that a given training stimulus induces different responses in the individual athlete. Coaches' experience even suggests the existence of two different groups, the responders and non-responders. This variability in training response is influenced by many different factors, such as genetic predisposition, mental status, pretraining status, training experience, age, and sex. Many athletes fail in trying to achieve an expected performance enhancement by copying published training regimes from other successful athletes. From a prevention perspective, the adjustment of training and exercises to each individual athlete is a major requirement. Training experience and age, for example, involve the danger of false or excessive loads that may lead to overuse syndromes.

Training regimes that are designed for an experienced athlete need to be adapted in terms of intensity, volume, duration of rest periods, and choice of exercise to serve for a training novice. Training for kids or adolescent athletes cannot just be a reduced expert level training regime but needs to consider the specific requirements of the maturing organism.

7.1.3.2 Specificity Principle

From a preventive point of view, specificity is one of the most important principles a coach should have in mind when designing a training program. A good example is strength training, where the exercise must be specific to the type of strength required and is therefore related to the particular demands of the event. The coach should have knowledge of the predominant types of muscular activity associated with his/her particular event, the movement pattern involved, and the type of strength required. Although specificity is important, it is necessary in every schedule to include exercises of a general nature. These exercises may not relate too closely to the movement of any athletic event, but they do give a balanced development and provide a strong base upon which highly specific exercise can be built [8].

7.2 Performance Components of Specific Sports Activities

7.2.1 Needs Analysis

Each event or discipline has specific requirements the athlete needs to fulfill. From a competitive and a preventive point of view, it is the basic motivation to know precisely what a discipline or event requires in terms of the different motoric skills (Fig. 7.1).

Is the athlete exposed to high external forces or is it necessary to create them internally by muscular action? Does he/she need to accelerate his body or just parts of it? What level of motoric skills is required and how long does the athlete have to execute the sports activity? Is endurance a limiting factor and is it more of aerobic nature or is the anaerobic portion dominant?

To match the principle of specificity, all these questions and many more need to be answered. The result of these efforts is a detailed and individual requirement profile for each athlete and each discipline or event. The individual component is given by the specific grade of aspiration that may reach from the recreational to elite level.

7.2.1.1 Endurance
According to the dominant energy contribution, we distinguish two basic components of endurance.

Aerobic and Anaerobic Endurance
Both components of endurance are important for the organism, and both are running simultaneously.

The first step in the sports-specific analysis focuses on determining the primary energy source used during the activity. The duration and intensity of the activity are the major determinants of the energy contribution.

The higher the intensity of the exercise and the shorter the duration of the activity, the more dominant anaerobic metabolism will get. When in contrast the intensity of exercise is reduced and duration of activity is elevated, aerobic metabolism will be the primary energy source.

7.2.1.2 Strength/Power
Force production, rate of force development, and time of appearance generally determine the different types of strength appearance.

Muscular action can be expressed statically or dynamically (concentric-eccentric phase) and take over an executing or stabilizing task.

- Muscular strength is the *maximum voluntary force* a muscle or muscle group can generate.
- Muscular power contains the explosive aspect of strength and speed – *explosive strength*.
- Muscular endurance is the ability of a muscle or a muscle group to repeatedly *sustain maximal or near-maximal forces*. According to Zintl [14], the level of force production must exceed 30 % 1 RM.

Maximum strength is the basis for all other subcategories of strength as it has determinant influence on all forms of force production.

7.2.1.3 Speed/Agility
In sports activities, speed can be expressed in many different forms. Speed reflects the ability to react on a specific signal or execute a given movement pattern in the shortest period of time. To distinguish speed from explosive strength, the aspect of low resistance has to be considered.

Speed requires a high level of neuromuscular activation and quality of movement coordination. In contrast to most of the other motoric skills, speed is highly determined by genetic factors such as muscle fiber composition.

In English and American literature, speed is commonly mentioned in combination with agility [2]. Agility describes the ability to change the body position efficiently and/or perform sports-specific movement skills with maximum speed, quality, and precision. Agility combines balance, coordination, and speed and trains or creates reflexes. These aspects make agility jump in the focus of preventive interventions in a great variety of sports activities.

Fig. 7.1 Motoric skills

7.2.1.4 Coordination

Zintl [14] defined coordination as the interaction of the CNS and the skeletal muscle system and is the basis of every human movement.

Athletes have access to multiple levels of motor control such as various sensory feedback systems. From a coordinative point of view, two fundamental phenotypes of movement exist [3]:

Closed-loop movements are characterized by a permanent comparison between the executed pattern via internal (sensory) and external feedback and the movement plan or program. On a lower level, closed loops are simple reflexes, but on a higher level, complex analysis and adjustment processes are enabled.

Open-loop movements require functioning feedback loops like the kinesthetic sensor system.

Open-loop movements like jumps or landings are controlled by feedforward mechanisms. The motoric action cannot be adjusted by feedback systems as it simply is too slow. The cognitive programming has to anticipate the movement with maximum accuracy to avoid unphysiological stress or strain. Training in open-loop situations requires the focus to be set on the quality of the movement.

Complex sports activities normally involve both closed- and open-loop situations.

7.2.1.5 Flexibility

In the context of sports, it means the ability to move a muscle or a muscle group over a complete or optimal range of motion. For a coach or an athlete, it is important to assess the influence of flexibility in their specific discipline to design the adequate training interventions.

In various disciplines like gymnastic or combat sports, flexibility is a performance-limiting factor. Training needs to focus on expanding the range of motion to a maximum level. Most of the other events use flexibility as preventive tool. The main focus is on providing suppleness of single muscles and thus ensures harmonic movement sequences.

Flexibility is dependent on many different factors besides the muscular system.

- Joint structure
- Rate of collagen fibers in the connective tissue
- Nervous system
- Pain
- Fatigue
- Age and sex

7.2.2   Preventive Approach

A more preventive perspective is primarily concerned with locating a possible dysbalance or unilateral load given by the structure of the sports or understanding the common sites of injury for a particular event. Tennis as a representative for court sports typically requires one strong racket arm. Together with the dominant arm, the whole-body side is exposed to unilateral loads when hitting the ball. Excessive or enduring training induces major dysbalances that may cause severe overstress situations in particular parts of the athlete's body.

Skiing is another good example for dominant parts of the body that are exposed to high stress and loads. High external forces have to be compensated mostly by the muscle groups of the lower limb and this was typically considered to be the limiting factor. Coaches' experience shows that many athletes complain about severe back pain as a result of emphasized strength training of the lower limb muscular system. An adjustment of the training philosophy following the ideas of functional training (6.3) helped reduce the back pain phenomenon significantly.

Coming back from an injury or training under the influence of pain may provoke motoric compensation patterns. Coaches need to detect situations like this and adjust the design of the training program to the specific needs of the athlete.

This indicates that the preventive approach of a need analysis should provide specific exercises that would strengthen particular joints or muscles with the purpose of preventing injuries or at least reducing their severity. This may improve the quality of performance by keeping the better athletes on the field of play longer.

> A systematic development of physical performance requires an individually adjusted training program that simultaneously considers both performance enhancement and preventive aspects [13]. This modern and synergetic approach is creating the conditions for a continuous and injury-free athlete development process.

7.3   Functional Training

Functional training is of great importance concerning prevention, rehabilitation, and performance optimization and is thus an inherent part of meaningful athletic training. It complies with the sports-specific demands and considers the athlete's individual preconditions. The goal in functional training is combining the respective athletic motor abilities to improve the efficiency of complex motion patterns within movement sequences as well as resolve compensation patterns and asymmetries in the athlete. Functional training is not a substitute for training of individual skills but an important addition. Certain performances are purposely trained isolated or functionally depending on training cycles.

A functional exercise program usually consists of free and unguided movements requiring larger amounts of coordination and sensomotor capability. Open-movement execution enables optimal selection of demands specific to the respective sports. It is the trainer's responsibility to analyze these and to create expedient exercise situations for the athlete.

7.3.1   The Sensomotor System

Our sensomotor system serves motion control and is composed of three levels. The complex information level for perception and transfer of information from the periphery uses external sensors and proprioception. The processing level is a central nervous capacity. Signals leading to a preferably precise coordinative output on the execution level are modulated here. Motor function is divided into two functional entities: postural and supporting motor functions (kinesthesia) and targeted motor functions (kinesis). Balancing these two entities is of great importance for the athlete's health and athletic performance.

7.3.2   Content and Goals of a Functional Training Program

The following elements are important parts of a functional training program and are specifically combined:

– Complex sensomotor tasks to improve neuronal adaptation and motor control
– Exercise content to improve muscular elasticity and sports-relevant agility
– Torso-stabalizing elements to increase core stability and segmental spine stability
– Exercises to improve the different strength forms and qualities, such as springiness, maximum force, and muscular endurance

The larger the overlap between the respective elements, the more complex the demand for the

Fig. 7.2 (**a**, **b**) Training example of a complex functional training

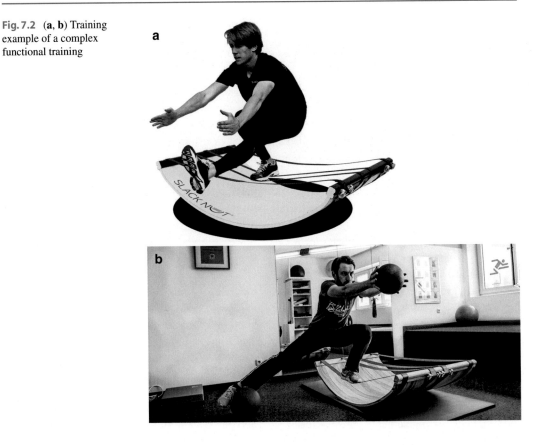

athlete. Examinations show mutual influence of athletic motor abilities on each other and provide evidence for a training physiological benefit [4, 12]. And this not only concerns sports-specific performance. Especially the combination of strength training and sensomotoric stimuli leads to biochemical stimulation and release of performance-enhancing growth hormones [6] (Fig. 7.2).

Examples for close and functional connection of strength, movability, and coordination: the athlete performs a strength-oriented exercise with maximum range of motion on an unstable surface made up of a teetering element and slack lines (Slack Nut®).

7.3.3   Effects and Benefits

Training programs with a large sensomotor proportion, such as exercises on an unstable surface or reactive jumping variations, have an extraordinary status in functional training.

Intensive stimulation of sensory feedback and its processing provably enhances muscle control. This is the key to a variety of protective and performance-enhancing adaptations. Inter- and intramuscular coordination is optimized by intensive stimulation of the sensory systems, improving whole-body coordination. On this basis, better joint control during stabilizing and kinetic tasks develops. Thus, improved dynamic standing stability can be established and falling risk can be likewise reduced [5].

Regular functional training with sensomotor components lowers the frequency of muscle and joint injuries by reduction of injury situations and training of the athlete's protective action ability. Various investigations show that approximately half of all injuries to the lower extremity could be prevented by specific sensomotoric training [4, 11, 13].

Apart from avoiding acute injuries, the athlete's long-term health must be considered in the training plan, and overuse through unvaried demands must be prevented. By intensive triggering of agonist muscles during sensomotor training units, muscular dysbalance is physiologically regulated and compensated. In this context, the positive influence on torso stability must also be mentioned, resulting from the complex motion linkup of "legs, torso, shoulders, and arms." The "core stability" improves force transmission onto the extremities and stabilizes the body's center of gravity.

Positive effects on strength development can also be demonstrated. Especially explosive strength in its initial phase is reinforced by sensomotoric connections. This promotes the athlete's speed, reactive motion sequences, and jumping performance [13], a benefit that is most relevant in acyclical and open play and motion forms and that improves the athlete's reactive and situative motion behavior [11].

Beside the described physiological adaptations, mental training aspects through complex and challenging movement patterns can be observed. Positive task solutions can support the psychological preparation for future strain and strengthen self-confidence [3].

Functional training programs with sensomotoric goals should be regularly varied to enable function expansion and to avoid habituation effects. Certain brain regions react to unpredictable and new motion tasks with increased release of neurotransmitters, such as dopamine and Munc-13. These substances have influence on our sensomotor system's processing and execution level [8]

7.4   Warm-Up and Cooldown

The goal of the warm-up period is to prepare the athlete both mentally and physically for exercise or competition. The warm-up should be an integral part of the prepractice or precompetition routine and does not require a large allocation of time. However, the warm-up should be given enough time to profit from the benefits.

The first step of a warm-up program can be of general nature and does not necessarily consist of exercises that are directly related to the specific event or discipline. The benefits of this *general warm-up phase* are increased blood flow to the exercising muscles, which increases muscle temperature and core body temperature. The increase in muscle and core body temperature has a significant positive effect on muscle strength and power and also improves reaction time and the rate of force development. Another collateral effect is the increase in blood temperature which increases the rate of oxygen resorption [7].

The second step of an adapted warm-up program integrates movements that closely simulate actions used during the activity the athlete is preparing for. The intensity level at the beginning is significantly reduced and rises to near competition level. Based on the first step, the benefits of this *specific warm-up phase* are the priming of the nervous system by increasing the rate and effectiveness of contraction and relaxation of both agonist and antagonist muscle groups and increases the discipline of specific elasticity and mobility of muscles, connective tissue, and joints [8, 10].

Warm-up may not just concentrate on physiological components but also integrate the mental aspects of the training or competition exercise. Mental preparation helps directing the athletes' concentration and assuring the status of complete vigilance.

"After activity means ahead of activity," following this saying, cooldown gets in the focus of further reflections about postexercise behavior.

Cooldown measures should support the organism to cope with the impacts of the previous activity and accelerate the regeneration.

Activities that expose the athlete to *high muscular loads* and stress should be compensated by tonus-regulating measures. Such can be low to moderate stretching exercises or thermic and/or physical applications.

Activities with *high metabolic requirements* should be compensated by low-intensity exercises that help rebalance the conditions in the used muscle cells. Additionally, various thermic and physical applications may help.

Similar to the warming up for a training or competition, cooldown also should integrate a mental component. Depending on the previous activity, the high alertness of the athlete and the induced neuromuscular irritability need to be calmed down.

> A structured and individually adapted program of warm-up and cooldown exercises is one major component in a holistic approach of long-term athlete development.

7.5 Physical Performance and Stress Components

Metabolic responses to stress and physical activity are extremely complex, involving many interacting variables. These multiple factors include endocrinological, physiological (cardiovascular and neuromuscular), biochemical, nutritional, and central nervous system (CNS) components, at a minimum.

The negative effects of overexercising can be physical and emotional. The athletes may see deterioration of their personal relationships as a result of their compulsions and frustrations or failure at work or school. Social isolation is common among those exercising compulsively as all available time is scheduled with physical activity, usually alone. The physical risks can be severe. An athlete can become unaware of dehydration or more seriously be affected by natural elements such as heat or cold and suffer heat stroke or frostbite, respectively.

Compulsive exercising can lead to insomnia, depression, fatigue, and anxiety. Physical side effects include muscular injuries and skeletal injuries such as shin splits, bone fractures, arthritis, or damage to cartilage and ligaments.

Too much exercise can lead to the release of excessive free radicals, which have been linked to cellular mutations and cancer. Females may cease menstruation and suffer from the so-called female athlete triad (anorexia/amenorrhea/osteoporosis). This means a serious health concern for active women and girls who are driven to excel in sports. According to the work of Amstrong et al. [1], in young military people, another significant problem combining acute weight loss with regular daily physical training may be a significant contributing risk factor for stress fracture injuries.

It is also known that repeated microtrauma of muscles and connective tissue eventually stimulates an autoimmune response, allergic reaction to foreign molecules, and inflammation. This leaves the body less able to handle infection. Also the continued glycogen depletion, leading to depleted energy stores in muscle and nerve cells, can also affect the autonomic nervous system, and prolonged excess levels of cortisol and other stress hormones can cause problems with the hormone regulatory system.

This potentially increased infection and immune response caused by amino acid imbalances resulting from prolonged training and/or an inadequate diet must be fully monitored in top level athletes.

The Table 7.2 resumes the fatigue-implicated factors that must be taken into account and corrected in sports training and performance.

The fatigue represents a homeostatic failure according to limitation model (Figs. 7.3 and 7.4) and needs an anticipation attitude in terms of correction.

Table 7.2 Particular factors in fatigue installation

Physiological
NCS muscle
Perfusion
Psychological
Insecurity, attention disturbances, work capacity, group problematics, etc.
Medical
Overtraining, infections, < immunity, trauma, etc.
Sports pedagogical
Wrong training methodology, insufficient rest, bad programs between competitions, wrong approach of individual aspects
Material – techniques
Sports material, competition and training camps, climate, diet, etc.

Fig. 7.3 Homeostatic
failure – limitations model

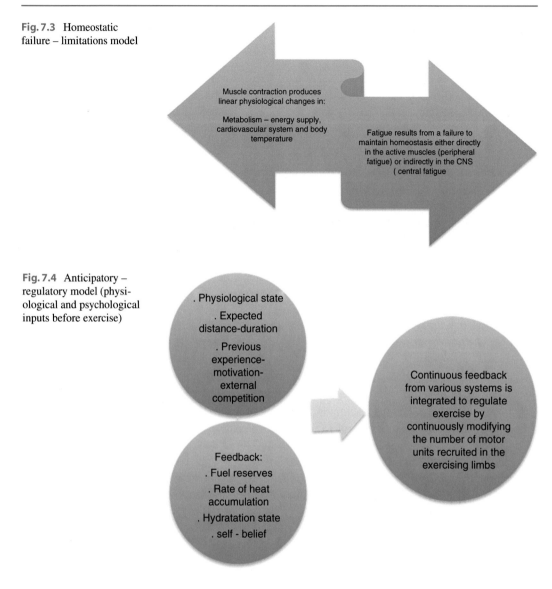

Fig. 7.4 Anticipatory –
regulatory model (physi-
ological and psychological
inputs before exercise)

7.6   Specific Adaptive Responses and Homeostasis

It is no secret among athletes that in order to improve performance they must work hard. However, hard training breaks you down and makes you weaker. It is rest that makes you stronger. Physiological improvement in sports only occurs during the rest period following hard training. This adaptation is in response to maximal loading of the cardiovascular and muscular systems and is accomplished by improving efficiency of the heart, increasing capillaries in the muscles, and increasing glycogen stores and mitochondrial enzyme systems within the muscle cells. During recovery periods, these systems build to greater levels to compensate for the stress that has been applied. If sufficient rest is not included in a training program, then regeneration cannot occur and performance plateaus. If this imbalance between excess training and inadequate rest persists, then performance will decline. The most common symptom is fatigue (Table 7.2). This may limit workouts and may be present at rest. The athlete may also become moody, become easily irritated, have altered sleep patterns, become depressed,

Fig. 7.5 Overtraining symptoms and recover strategies

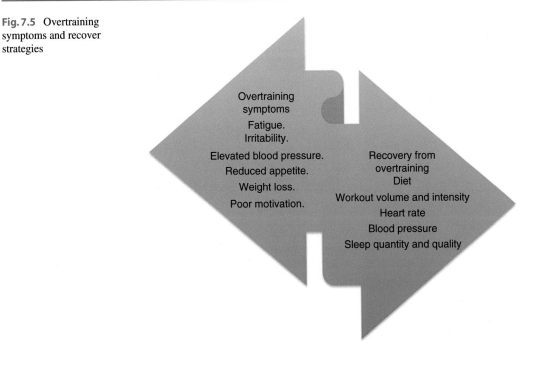

or lose the competitive desire and enthusiasm for the sports. Some will report decreased appetite and weight loss. Physical overtraining symptoms (Fig. 7.5) include persistent muscular soreness, increased frequency of viral illnesses, and increased incidence of injuries which need appropriate recovery measures. Between these measures, hyperbaric oxygen could help recover after vigorous exercise [1]. Oxygen breathing after exercise may provide positive psychological effect, muscle recovery sensation, and lower lactate and heart rate to athletes.

In resume, the goal is to achieve an appropriate balance between training and competition stresses and recovery in order to maximize the performance of athletes. A wide range of recovery modalities are now used as integral parts of the training programs of elite athletes to help attain this balance.

Even without considering the above limitations, there is no substantial scientific evidence to support the use of the recovery modalities reviewed to enhance the between-training session recovery of elite athletes [1]. Modalities as massage, active recovery, cryotherapy, contrast temperature water immersion therapy, hyperbaric oxygen therapy, nonsteroidal anti-inflammatory

drugs, compression garments, stretching, electromyostimulation, and combination modalities need further investigation regarding efficacy as recovery modalities for elite athletes. Other potentially important factors associated with recovery, such as the rate of postexercise glycogen synthesis and the role of inflammation in the recovery and adaptation process, also need to be considered in this future assessment.

7.7   Risk Management

Sports on each level of ambition, in every peculiarity and no matter what event, has a major fascinating impact on the athletes. To perform a carved turn in skiing and feel the dynamics of external forces or hitting the ball with the sweet spot of the racket and smashing it precisely in back corner of the tennis court or even running a certain distance in your personal best time and feel as if you would fly over the ground is something you cannot buy but just experience. Where else can you push your personal limits and get positive acknowledgment even by people who don't share the same passion.

Fig. 7.6 Model of hazard evolution

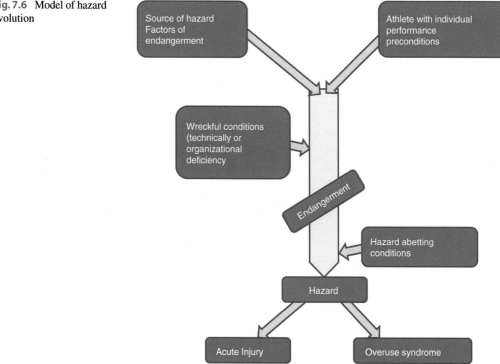

Unfortunately, this fascination simultaneously can turn into the main risk factor.

The high forces in the carved turn, the ambition to hit that short stop ball with the last of your strength, or to go "all in" at your last training session of the week can easily turn fascination into disaster.

This shows that discussing the role of prevention in sports needs deal with that ambivalence.

Personal ambition is one of the major components that can influence that conflict. Each individual either coach or athlete needs to meticulously determine the requirements of the event or discipline they are preparing for and assess the individual performance level of the motoric skills they figured out in that process. That next important step is to contrast both aspects and then formulate the suitable goals. Based on that result, the adapted training program can be designed. Personal ambition needs then to be adjusted accordingly.

One thing that needs to be kept in mind is that besides the individual ambition, external pressure may distort this demand – precondition – analysis.

Effective prevention programs therefore need to individually analyze and manage the above-described risk factors [10].

A German employer mutual insurance association VBG [9] developed a model of hazard evolution that can also be applied in the context of sports. Using models like this may be helpful not only to explain the genesis of an actual hazard out of an obviously innocuous situation but also to figure out where the starting points for prevention considerations are located (Fig. 7.6).

To illustrate the model endurance, running that – at a first glance – isn't quite hazardous may serve as a valuable example.

Source of hazard:
Execute the action of running
Factors of endangerment:
Biomechanical stress on active and connective tissue
Athlete's individual performance preconditions:
Age, sex, training experience, performance level, running skills, etc.

Wreckful conditions:

Technically – inadequate running equipment (shoes)

Organizational: inadequate training regimen, training group or partners (level of performance), etc.

⇒ *Resulting endangerment*:

Possible overuse/overstress

Hazard-abetting conditions:

Lifestyle, diet, fatigue, running style, goal setting, etc.

⇒ *Hazard*:

Emerge of the acute overuse

⇒ *Consequences*:

Injury: fatigue fracture

Overuse syndrome: inflammation of the Achilles tendon

Following this hazard evolution model clearly highlights where prevention programs could intervene in this particular case.

Attenuating the risk-enhancing factors would indicate to supply the proper running equipment to avoid unphysiological stress. Another measure should focus on adjusting the training regime and even the selection of the training partner of the athlete.

Additionally, the lifestyle, diet, and the goal setting of the runner need to be adjusted to individual performance preconditions, time frame, and personal ambitions.

> In modern lifestyle, there are, superficially seen, various options to pass responsibility on to external actuators such as personal trainers, guides, instructors, event organizers, and even wearable sensors and their corresponding "Apps."

> This shift of responsibility is actually not an effective prevention but just an ostensive protection.

References

1. Barnett A (2006) Using recovery modalities between training sessions in elite athletes: does it help? Sports Med 36(9):781–796
2. Cardinale M, Newton R, Nosaka K (2011) Strength and conditioning – biological principles and practical applications. Wiley-Blackwell, Oxford
3. Diemer F, Sutor V (2008) Praxis der Trainingstherapie. Georg Thieme Verlag KG, Stuttgart
4. Emery CA, Meeuwisse WH (2010) The effectiveness of a neuromuscular prevention strategy to reduce injuries in youth soccer: a cluster-randomized controlled trial. Br J Sports Med 44:555–562
5. Gisler-Hofmann T (2008) Plastizität und Training der sensomotorischen Systeme; Schweizerische Zeitschrift für. Sportmedizin und Sporttraumatologie 56(4):137–149
6. Haas CT, Turbanski S, Schwed M, Schmiedbleicher D (2006) Neuronale Korrelate apparativ gestützter Trainingsformen. In: Witte K, Edelmann-Nusser J, Sabo A, Moritz EF (eds) Sporttechnologie zwischen Theorie und Praxis. Shaker, Aachen, pp 37–48
7. Herman K, Barton C, Malliaras P, Morrissey D (2012) The effectiveness of neuromuscular warm-up strategies, that require no additional equipment, for preventing lower limb injuries during sports participation: a systematic review. BMC Med 10:75
8. Hoffmann J (2014) Physiological aspects of sport training and performance, 2nd edn. Human Kinetics, Champaign
9. Schubert K, Littinski R, Ludborzs B (1997) Sicherheits-Audits: Effizienzsteigerung im Arbeits- und Gesundheitsschutz. Universum Verlag, Wiesbaden
10. Laursen JB, Bertelsen DM, Andersen LB (2014) The effectiveness of exercise interventions to prevent sports injuries: a systematic review and meta-analysis of randomised controlled trials. Br J Sports Med 48:871–877
11. Lindblom H, Waldén M, Carlfjord S, Hägglund M (2014) Implementation of a neuromuscular training programme in female adolescent football: 3-year follow-up study after a randomised controlled trial. Br J Sports Med 48:1425–1430
12. Ringel S (2015) Effekte eines sensomotorischen Trainings auf der SlackNut auf die Kraft und Balance bei Fußballspielern: Eine prospektive randomisierte kontrollierte Studie. Univ Masterthesis, Köln
13. Taube W (2012) Neurophysiological adaptations in response to balance training. DtschZ Sportmed 63: 273–277
14. Zintl F (1990) Ausdauertraining. BLV Sportwissen, München

Specific Aspects of Throwing Sports in Recreational and Competitive Sport

8

Joris R. Lansdaal, Michel P.J. van den Bekerom, Ann M.J. Cools, Val Jones, Nicolas Lefevre, and Elvire Servien

J.R. Lansdaal, MD (✉)
Department of Orthopaedic Surgery,
Central Military Hospital/University Medical Centre,
Utrecht, The Netherlands
e-mail: jrlansdaal@gmail.com

M.P.J. van den Bekerom, MD
Department of Orthopaedic Surgery,
Onze Lieve Vrouwe Gasthuis,
Amsterdam, The Netherlands
e-mail: Bekerom@gmail.com

A.M.J. Cools, PhD, PT
Department of Rehabilitation Sciences and
Physiotherapy, Ghent University, Ghent, Belgium
e-mail: ann.cools@ugent.be

V. Jones, PT
Sheffield Shoulder and Elbow Unit, Department of
Orthopaedic, Northern General Hospital, Sheffield,
United Kingdom
e-mail: valjones2305@googlemail.com

N. Lefevre, MD
Department of Orthopaedic Surgery,
Clinique du Sport Paris, Paris 75005, France

Department of Orthopaedic Surgery,
Institut de l'Appareil Locomoteur Nollet,
Paris 75017, France
e-mail: docteurlefevre@sfr.fr

E. Servien, MD, PhD
Department of Orthopaedic Surgery,
Hopital de la Croix-Rousse, Centre Albert Trillat,
Lyon University, 69004 Lyon, France
e-mail: elvire.servien@chu-lyon.fr

Key Points
1. For the development of prevention strategies of injuries in throwing sports, it is necessary to understand the throwing mechanism.
2. The risk factor with the strongest correlation to elbow and shoulder injury is the amount of pitching, hitting or throwing.
3. In prevention of injury in throwing sport it is vital to concentrate on the whole "Kinetic chain".
4. Maintaining range of motion, optimising upper extremity and scapula strength and endurance and improving neuromuscular control, are important factors in injury prevention of the throwing athlete.
5. Most research in throwing sports is on baseball pitching, this can not directly be extrapolated to other throwing sports.

8.1 Physical and Psychological Aspects in Throwing Sports in Recreational and Competitive Sports

The throwing athlete is at risk to develop upper extremity injuries as a result of the repetitive high forces during the act of throwing [1]. Most

© ESSKA 2016
H.O. Mayr, S. Zaffagnini (eds.), *Prevention of Injuries and Overuse in Sports: Directory for Physicians, Physiotherapists, Sport Scientists and Coaches*, DOI 10.1007/978-3-662-47706-9_8

research on throwing sports is done within the field of baseball so most evidence and knowledge refer to this and are not directly extrapolated to other throwing sports. Half of the injuries occurring in baseball involve the upper extremity and mainly the shoulder and the elbow. Pitchers at high school, collegiate and professional level are most prone to injury compared to the position players and do more often result in surgical intervention [2]. Besides the direct trauma injury, most of the pitching injuries develop from overuse and, very likely, from over a time span of many seasons.

The consequences of upper extremity injury are diverse and vary in impact. The injury can result in a prolonged leave form sport, having consequences for the individual and also for the team. The injury can develop problems in daily life, work and school as described by Register-Mihalik et al. [3]. It will also result in additional costs, not only for the athlete and his or her family but also for schools and social security. Eventually surgery can be needed, taking an additional long recovery period. For example, the ulnar collateral ligament reconstruction, which is one of the most often-preformed elbow surgeries, will take a person a recovery time from 12 to 18 months. All this can bring a future in performing throwing sports at risk.

When we take all these consequences in consideration, there is an urgent need for prevention of upper extremity injury and research in this field. Focus has to be on recreational and professional adolescents, elderly, but has to start with youth. Proper youth training has most chance to preventing injury and also has the additional benefit that preventive measured implementations may help to reduce injuries later on or at higher playing levels.

Many authors suggest three potential approaches to prevent upper extremity injury in throwing sports [4, 5]:

1. Regulation of unsafe participating factors,
2. Exercise intervention to modify suboptimal physical characteristics
3. Instructional intervention to correct improper throwing techniques

The regulation of unsafe participating factors including number of pitches over a single outing and over the season is classified and regulated by, e.g. Little League™ Baseball and the USA Baseball Medical Safety Board. They recommend age-specific pitch count limits and advise rest periods for the prevention of overuse.

Exercise intervention to modify suboptimal physical characteristics and instructional intervention to correct improper throwing techniques are discussed in the next paragraphs.

8.2 General Training Aspects in Throwing Sports

The act of throwing requires good coordination in one controlled motion from the foot to the hand. This is referred to as the "kinetic chain". For the principle of the kinetic chain to work, the body has to effectively transmit the generated energy from the lower body into the upper body ending eventually in adequate ball velocity and direction. The accuracy and velocity are determined by the effectiveness of the muscle sequence and timing. Body rotation and scapular position are very important factors in the kinetic chain [6]. Physical conditions that influence the kinetic chain, and especially the core, will interact with the more distal part of the body and can lead to pathological shoulder and elbow conditions.

The throwing motion is divided into six phases (Fig. 8.1): (1) windup, (2) cocking, (3) late cocking, (4) acceleration, (5) deceleration and (6) follow-through. Because specific injuries develop in certain phases, it is important to know in what phase the injury happened. The throwing motion takes approximately 2 s to complete.

The first 1.5 s consists of the first thee phases, and the following phase, acceleration, takes only 0.05 s. In this phase the largest change in rotation and also the greatest angular velocities occur. Because of this most injuries happen during the acceleration phase. The last two phases take about 0.35 s.

The shoulder is exposed to very high forces during the throwing motion. The speed of a pitched ball can reach up to 144 km/h bringing up angular velocities in the shoulder up to 7000°/s, and a shoulder is exposed to distractive (950 N), compressive (1090 N) and shear (400 N) forces.

The shoulder and elbow are working through dynamic as well as static stabilisers. The

Fig. 8.1 (1) Windup, (2) cocking, (3) late cocking, (4) acceleration, (5) deceleration and (6) follow-through

dynamic stabilisers of the glenohumeral joint are the rotator cuff, the long lead of the biceps and the scapulothoracic muscles. The static stabilisers are the glenohumeral capsule, the labrum and the bone. The dynamic stabilisers of the ulnoradial-humeral joint are the flexor, pronator and extensor muscles. The static stabilisers are the joint capsule and the lateral and medial ligaments and the bone. This is mainly in full extension or flexion and less in midrange. These structures can be disrupted by a traumatic event or by overuse.

There is a balance in the shoulder and elbow between stability and mobility. The shoulder needs to be mobile enough to reach extreme positions, for example, maximum exorotation, to achieve maximum velocity. On the other hand it needs to be stable enough to keep the humerus in the glenoid in place and keep a functioning kinetic chain. With every pitch there is a tremendous load on the soft tissue of the shoulder and elbow, which makes it vulnerable to injury. This balance in stability-mobility can be altered and eventually lead to injury in high repetitions and increased demands.

8.3 Specific Preventive Activities for Shoulder and Shoulder Girdle

Based on recent prospective studies, several risk factors for shoulder overuse injury have been defined in a variety of sports, including baseball, handball, rugby and tennis. In general these risk factors can be divided into three categories: (1) loss of rotational range of motion, in particular loss or internal rotation and/or total range of motion; (2) decreased rotator cuff strength, in particular in the external rotators; and (3) scapular dyskinesis. In view of these risk factors, preventive programmes should focus on restoration or maintenance of normal values of these variables, using the appropriate strengthening and stretching exercises.

8.3.1 Glenohumeral Range of Motion

With respect to range of motion, loss of internal range of motion is known to be a risk factor for chronic shoulder pain. Therefore it is advised that side differences in internal rotation ROM should be less than 18°, and the difference in total range of motion should not be more than 5°. The assessment of the ROM into rotation of the shoulder can be measured with a goniometer or an inclinometer and in many positions of the body and the shoulder. A comprehensive reliability study [31] showed high to excellent inter- and intra-tester reliability for a variety of test positions and equipment. Based on the results of this study and in view of optimal standardisation of body and shoulder position, the authors advise the following procedure: the patient is supine with the shoulder in the frontal plane and the elbow flexed 90°. The upper arm should be horizontal, if needed (e.g. in case the patient has protracted shoulders or a thoracic kyphosis), the arm is supported by a towel to reach the horizontal position. For internal rotation the examiner palpates the spine of the scapula and the coracoid. The inclinometer is aligned with the forearm (olecranon and styloid process of the ulna), and the shoulder is moved into internal rotation (Fig. 8.2). The movement reaches its endpoint when the coracoid tends to move against the palpating thumb. For external rotation the fixating hand is gently put over the shoulder top, and the shoulder is moved into external rotation, aligning the inclinometer with the forearm.

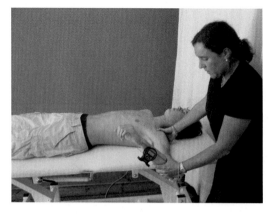

Fig. 8.2 Measurement of internal rotation of the shoulder, using a digital inclinometer

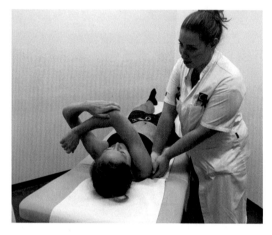

Fig. 8.3 Cross body stretch

Given the evidenced impact of posterior shoulder tightness on shoulder kinematics, increasing posterior shoulder flexibility is advisory when mobility deficits exceed the limits associated with increased injury risk. Both the cross body stretch (Fig. 8.3) and the sleeper stretch (Fig. 8.4) can be recommended to decrease posterior shoulder tightness. It was shown that a 6-week daily sleeper stretch programme (3 reps of 30 s) is able to significantly increase the acromiohumeral distance in the dominant shoulder of healthy overhead athletes with GIRD. Additional joint mobilisation performed by a physiotherapist has a small but non-significant advantage over a home stretching programme alone. No difference in mobility gain was seen after angular (sleeper stretch and horizontal adduction stretch) and non-angular (dorsal and caudal humeral head glides) joint mobilisation by a physiotherapist. Muscle energy techniques (hold-

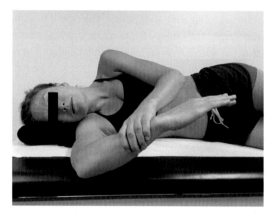

Fig. 8.4 Sleeper stretch

relax) during the sleeper stretch and the horizontal adduction stretch have proven useful to immediately increase internal rotation range of motion.

8.3.2 Rotator Cuff Strength

Regarding rotator cuff strength, it is generally recognised that overhead athletes often exhibit sport-specific adaptations leading to a relative decrease of the strength of the external rotators. Absolute side differences as well as muscle balance ratio between external and internal rotators have been found to be altered. In general, with respect to cut-off values distinguishing a healthy shoulder from a shoulder at risk, an isokinetic ER/IR ratio of 66 % or an isometric ER/IR ratio of 75 % (when measured in neutral position) is advised [31], with a general rotator cuff strength increase of 10 % of the dominant throwing side compared to the nondominant side. Recently, focus has shifted from isometric or concentric to eccentric muscle strength of the rotator cuff. In particular the eccentric strength of the external rotators is of interest. These muscles function as a decelerator mechanism during powerful throwing, serving or smashing.

In view of the importance of eccentric rotator cuff strength in relation to injury-free overhead throwing or serving, it is imperative that strength is assessed on a regular basis in healthy as well as injured players. With respect to the isometric strength measurements, hand-held dynamometry (HHD) has attracted more and more interest during the last years due to the more practical, less expensive and user-friendly advantages over the more advanced and expensive isokinetic devices (Fig. 8.5).

Recently, a new testing protocol was published, showing that the use of a hand-held dynamometer (HHD) measuring eccentric external rotator strength has excellent intra-tester (ICC = 0.88) and good inter-tester (ICC = 0.71) reliability, as well as concurrent validity (compared to an isokinetic device, Pearson correlation = 0.78). During the procedure, the patient is seated with gentle support of the arm by the tester; the tester brings the shoulder from 90° abduction–90° external rotation (throwing position) to 90° abduction–0° external rotation, loading the external rotators eccentrically (Fig. 8.4). A large normative database on 200 overhead athletes (volleyball, tennis and handball) was recently set up and shows an average normalised eccentric external rotator strength (N/kg) of approximately 2, with significant side differences in favour of

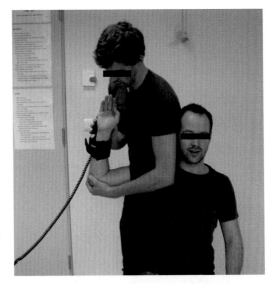

Fig. 8.5 Eccentric testing protocol, using an HHD

Fig. 8.6 Eccentric exercise for the external rotators in an abducted position

the dominant sides and significant higher values for handball and tennis compared to volleyball.

Numerous exercises have been described to strengthen the rotator cuff muscles, including concentric, isometric, eccentric and plyometric exercises. In view of the eccentric component of the function of the external rotators, the sport-specific exercises for overhead athlete players should focus on three areas:

1. Accentuate the eccentric phase and "avoid" concentric phase, in order to load the muscles based on their eccentric capacity. Figure 8.6 shows an example of an eccentric exercise for the external rotators in general in an abducted position.

2. Slow exercises for absolute strength and fast exercises for endurance and plyometric capacity. Endurance and plyometric capacity may be exercised using weight balls, in which the patient is instructed to "catch" the ball (Fig. 8.7), as described by Ellenbecker and Cools.

3. Exercise highlighting the "stretch-shortening cycle" of throwing. Specific devices can be used to train the stretch-shortening cycle, such as XCO® trainer (Fig. 8.8).

8.3.3 Scapular Dyskinesis

Evidence supporting cut-off values for prevention of injury or return to play after injury with respect to scapular function is scarce. A number of studies used visual observation as a criterion, whereas others provide objective data on healthy athletes as a

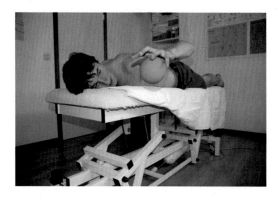

Fig. 8.7 "Catching" exercise, using a Plyoball

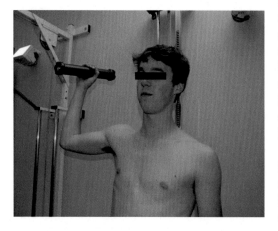

Fig. 8.8 Stretch-shortening cycle exercise, using the XCO© trainer

reference base for return to play. In general, visual observation is performed either by using the yes/no method (scapular dyskinesis or not), a method proven to be reliable and valid, on the condition that the examiner/therapist is educated in a standardised manner, or by categorising the scapular dysfunction into different types, based on the specific position of the scapula. However, the latter method was shown to have acceptable intra-rater, but low inter-rater reliability. A statement saying that scapular behaviour should be symmetrical in overhead athletes is not supported by research data, on the contrary. In volleyball, as well as in handball players, asymmetry was found in resting scapular posture. Therefore clinicians should be aware that some degree of scapular asymmetry may be normal in some athletes. It should not be considered automatically as a pathological sign but rather an adaptation to sports practice and extensive use of the upper limb.

Several studies measured scapular upward inclination in healthy overhead athletes. These data may be used as a reference base and cut-off values for correct scapular positioning in several elevation angles. In general, a large variety is found in scapular upward inclination in the mid range of motion (probably due to a large variation between individuals); however, in full elevation, most studies suggest that upward inclination should be at least 45–55°.

For the scapular muscles, proper inter- and intramuscular balance should be assessed. Isokinetic ratio protraction/retraction is shown to be 100 % in a healthy population, with slight changes in overhead athletes, in case of throwing athletes in favour of the protractors. In bilateral sports (swimming, rowing, gymnastics), there should be no side differences in scapular muscle strength, and in one-handed overhead sports, an increase of scapular muscle strength of 10 % is advised on the dominant side. In particular, the lower trapezius and serratus anterior should receive special attention, since these muscles are shown to be susceptible to weakness in injured athletes.

Once deficits and imbalances in scapular behaviour are assessed, an intervention programme to restore flexibility and muscle performance needs to be installed. Recently, a science-based clinical reasoning algorithm was published guiding the clinician into the different steps and progression. The main goals are (a) to restore flexibility of the surrounding soft tissue of the scapula, in particular the pectoralis minor, levator scapulae, rhomboid and posterior shoulder structures, and (b) to increase scapular muscle performance around the scapula, focusing on either muscle control and inter- and intramuscular coordination or muscle strength and balance. Exercises to restore scapular muscle balance have been shown to increase isokinetic protraction and retraction strength, increase external rotator strength of the shoulder and alter EMG activity of the scapular muscles in favour of efficient muscle recruitment during a loaded elevation task. In general focus should be put on training the lower and middle portions of the trapezius (Fig. 8.9) and training the scapular muscles in higher elevation angles, relevant to the specific sport (Fig. 8.10).

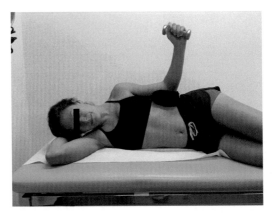

Fig. 8.9 Exercise for lower and middle portions of the trapezius

8.4 Specific Preventive Activities for Elbow

Elbow injuries occur throughout many but most in overhead sporting activities [21, 22]. The increase in the number of participants in sports as baseball, tennis, football, javelin throwing and volleyball has probably resulted in a rise in the incidence of elbow injuries in recent decades.

The incidence of shoulder and elbow injury in baseball is related to the number of years of participation, as well as to the age of athlete; elbow pain is reported in 20 % of the 8–12-year-old group, 45 % of the 13–14-year-old group and 58 % of high school and college athletes [17, 18]. The majority can be classified as either overuse or acute traumatic injuries.

The forces commonly encountered at the elbow in sports are best documented in the overhead-throwing athlete. The overhead-throwing motion can be divided in six phases as described earlier: (1) windup, (2) early cocking, (3) late cocking, (4) acceleration, (5) deceleration and (6) follow-through [5, 6]. This motion especially during the late cocking and early acceleration phase and in the follow-through phase generates valgus and extension forces. Valgus stress can reach 64 Nm, with compression forces in the radiocapitellar articulation as high as 500 N and extension velocities in excess of 3000° per second. These combined forces have been termed

Fig. 8.10 Exercise for the scapular muscles in higher elevation angles

valgus extension overload (VEO) and are believed to be responsible for most injury patterns seen in the throwing athlete [7]. These forces consist of lateral compressive forces (radial head/capitellum), large tensile forces that occur in the medial structures (UCL, pronator/flexor muscle mass, medial epicondyle) and shear stresses that occur in the posterior compartment (olecranon tip, olecranon fossa). These forces can finally result in attenuation or rupture of the UCL, flexor pronator tendinopathy, formation of olecranon tip osteophytes or loose bodies, olecranon stress fractures and osteochondral defects of the capitellum [8, 14, 15]. Most scientific research focusing on preventive measures for overuse elbow problems is performed in baseball players (by Glenn Fleisig and colleagues) so the advices discussed below are mostly based on baseball players.

8.4.1 Quantity

The risk factor with the strongest correlation to elbow injury is the amount of pitching. Based on the research of Olsen et al., there they concluded that averaging more than 80 pitches per game almost quadrupled the chance of surgery (OR 3.83) and pitching competitively more than 8 months per year increased the odds of surgery by fivefold (OR 5.05). A pitcher who regularly pitched with his arm fatigued was 36 times as likely to have an injury which required surgery (OR 36.18). Little League and high school pitchers who also pitched for travel teams or showcases had increased risk of elbow injuries. Studies show that it is not uncommon for youth pitchers to play 70 games or more per year, often playing more than 8 consecutive months [23]. The rise in overuse elbow injuries in young pitchers corresponds with the extended competitive baseball of recent times.

8.4.2 Biomechanics

Maximum values of internal rotation torque of the shoulder and varus torque of the elbow are produced near the time of maximum external rotation to decelerate shoulder external rotation, prevent elbow valgus opening and initiate shoulder internal rotation. Tension in the UCL absorbs about half the varus torque in this position. Pitchers at the youth level have greater inconsistency in their biomechanics from pitch to pitch [12]. Improper pitching mechanics can lead to increased elbow varus torque and consequently increased risk of elbow injury.

Changes in other parts influence the elbow biomechanics during the throwing motion: trunk range of motion, late arm rotation, excessive shoulder external rotation, passive total shoulder rotation, excessive elbow flexion and improper shoulder abduction and trunk tilt [9, 10, 13, 16]. The study of Huang and colleagues demonstrated that youth baseball players with a history of elbow pain threw with a more extended elbow at maximum shoulder external rotation and greater lateral trunk tilt at ball release [11]. But there is not known if the pitchers with an injury history demonstrated the error prior to the time of injury or if

the error developed after the injury. Improving elbow and body biomechanics can increase a pitcher's chance of staying healthy [24].

8.4.3 Technique

For several years many baseball experts stated that curveballs result in pain and injury in youth baseball pitchers. The theory is that to spin a ball to cause a break, the pitcher must put his arm in a position that increases strain on the elbow [24]. However recent research showed that elbow varus torque was less in curveballs compared to fastballs. Fleisig and colleagues found no difference in varus torque between fastballs and curveballs [25].

Research focusing on the relationship between throwing curveballs at a young age and elbow pain or injury does not show a relationship between age and throwing curveballs on one hand and the risk of injury on the other hand.

8.5 Specific Preventive Activities for Trunk, Spine and Lower Extremity

In order to generate an efficient throwing motion resulting in maximum precision and velocity, the lower and upper body has to work in a coordinated and synchronised way.

It is vital not only to concentrate on the upper limb to prevent injury but on the whole kinetic chain.

Injuries or adaptations in remote areas of the chain can cause problems not only locally but also distally, as joints such as the elbow compensate for the lack of force production and energy delivery through more proximal links. The stance leg hip extensors are responsible for controlling and maintaining hip flexion by eccentrically and isometrically contracting. For this reason, weak hip abductors or quadriceps in the stance leg create an unstable platform for the more distal components of the kinetic chain [24]. The shoulder and elbow have to work extra hard to compensate and maintain throwing precision and velocity. Kibler and Chandler [26] calculated a 20 % reduction in kinetic energy delivered from the hip and trunk to the upper limb. This requires a 34 % increase in rotational velocity of the arm, to impart the same amount of force to the hand. Hannan et al.'s study [27] has shown a link between lower-limb balance deficits in throwers with medial elbow ligament injuries and healthy controls. These balance deficits disappear following a 3-month throwers rehab programme including the trunk and the lower limb [27].

If we look at the difference of pitching in various levels, one of the cornerstones is the timing of trunk rotation. A high-level pitcher will show a delayed trunk rotation when compared with a less experienced pitcher. The high-level and professional pitchers also have less humeral internal rotation torque when compared with collegiate pitchers. This raised the idea that early rotation of the torso is inefficient and can cause injury. It is assumed that starting trunk rotation before the humerus and scapula are in the correct position results in excessive horizontal abduction and hyperangulation. Also lack of internal rotation of the stance leg hip can lead to premature opening and forward rotation of the pelvis, which can result in more distal demands of the kinetic chain.

Also the use of excessive contralateral trunk lean (Fig. 8.11) results in more velocity and also increased joint loading.

For all of these reasons, leg and trunk exercises involving sport-specific activation patterns can be implemented, enhancing the transfer of kinetic energy from proximal to distal segments.

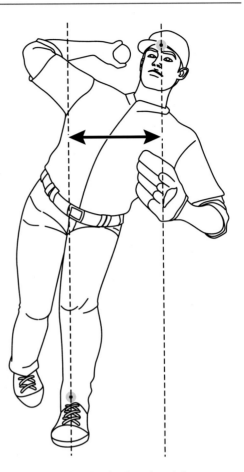

Fig. 8.11 A pitcher showing lateral trunk lean

8.6 Prevention of Overload and Reducing the Risk of Injury in High-Performance Throwing Athlete

The overhead-throwing athlete is extremely challenging to manage, as the repetitive microtrauma placed on the upper-limb joints during throwing motion can exceed the physiological limits of the surrounding tissues. It is therefore imperative to emphasise the preventative elements in caring for these athletes. Injury prevention is a process whereby an athlete is screened through a variety of tests to identify any potential problems within the musculoskeletal system. Once these problems are identified, practices are put into place to eradicate or reduce their possible impact, including educating athletes regarding possible warning signs of injury.

Numerous risk factors in throwing athletes have been identified, defined as either non-modifiable or modifiable. Non-modifiable factors include age, body mass index, height, coaching habits and satisfaction with performance. Modifiable factors include altered throwing mechanics, frequency and volume of throwing activities and individual physical characteristics including decreased flexibility, fatigue, weakness of upper and lower limbs, rotator cuff imbalance or altered core stability. Strategies to address modifiable risk factors are discussed below [19].

8.6.1 Maintaining Range of Motion

Shoulder flexibility is assessed and addressed, as loss of total shoulder rotational range or glenohumeral internal rotation deficit (GIRD) has been shown to place strain on the shoulder and medial elbow structures during throwing [4]. Currently there is little consensus regarding the tissues responsible for this change; some authors believe postero-inferior glenohumeral capsular tightening and shrinkage are responsible, whilst others attribute the changes to adaptive humeral head changes seen in long-term throwers [4, 20]. For this capsular tightening group, stretches such as a sleeper stretch are thought to be effective in addressing the restriction. However, some authors consider that restriction of rotational range may be a problem of dynamic control rather than of capsular origin [28]. For this group stretching may not be indicated, and the problem should be addressed by performing rotator cuff control exercises through range [28]. It is, therefore, essential that the individual is carefully assessed to ensure that any deficit is managed appropriately.

8.6.2 Maintenance of Upper Extremity and Scapula Strength

The act of throwing is challenging for the dynamic stabilisers of the entire upper limb, including those of the scapula. Strength deficits around the shoulder and scapula are significantly associated with increased injury rates in the shoulder and elbow in throwing athletes. Strengthening programmes such as the Throwers 10 and the Pitcher's Baseball Bat Programme have been designed, from EMG evidence, to illicit muscular activity most needed to provide upper-limb dynamic stability, and have been demonstrated to increase throwing velocity, following a 6-week programme.

However, it should be noted that muscle group strength ratios are sport specific. In some overhead activities such as volleyball and tennis, high elbow extensor to flexor ratio is seen, whereas in activities such as judo, there is an almost equal ratio of elbow extensors to flexors. This should be borne in mind when designing individual rehabilitation programmes.

8.6.3 Dynamic Stability and Neuromuscular Control

Emphasis is also placed on exercises improving endurance and neuromuscular control of the elbow complex. Loss of kinaesthetic awareness of the upper extremity has been shown to decrease proprioceptive accuracy in throwers [29]. Studies show a decrease in neuromuscular control, kinaesthetic detection strength and throwing accuracy is associated with muscular fatigue [29]; therefore, exercises, including multiple sets, to promote endurance are a key component in throwing athlete training regimes.

Fatigue has been associated with an increased risk of injury to the shoulder and elbow in throwers, and therefore, recognition of fatigue as an early warning sign by both players and coaching staff and incorporation of rest into training programmes have been advocated.

Proprioceptive neuromuscular facilitation, plyometric exercises, rhythmic stabilisation drills and open and closed kinetic chain activities, which promote co-contraction and mimic functional positions with joint approximation, are also necessary for overhead athletes to reduce their chance of injury.

8.6.4 Impact Work

As well as producing force, training should develop the overhead athlete to accept force. Sport requires an athlete to reduce and absorb external forces often at high speeds in three dimensions, in an unpredictable environment. Training programmes need to incorporate rapid acceleration and deceleration, changes of direction and direct impact. For individuals who wish to return to contact sports, e.g. rugby, it is vital to address impact work. Previous studies have shown that increased muscle activation patterns of the elbow and wrist during forward falls increase the transition of force shock waves through the forearm. With practice, individuals can select the upper extremity posture during forward falls, allowing the athlete to minimise the effects of impact.

8.6.5 In-Season Training

The aim of training during the competitive season is to maintain gains made in the off season in strength and conditioning, as a long competitive season can result in a decline in physical performance. Maintenance programmes should focus on strength and stability whilst adjusting for the workload of a competitive season.

To avoid overload in younger baseball pitchers, guidelines from the USA recommend a limit on game and yearly pitch counts, as higher single-game and cumulative game pitch counts are associated with an increased incidence of upper-limb pain. The guidelines also recommend the avoidance of yearlong participation in baseball and advise rest periods throughout the year.

Throwing style may also need addressing. Pitching from a mound creates a mechanical advantage with increased efficiency, compared with throwing from flat ground, thereby throwing from a mound should precede throwing from other types of surfaces.

8.6.6 Off Season

This is a valuable time to treat previous injuries and prepare for the upcoming season. At the conclusion of a competitive season, athletes should remain physically active whilst taking time away from repetitive overload. The remainder of the off season is to build a baseline of strength, endurance, neuromuscular control and flexibility before starting sport-specific drills.

All athletes should be educated on proper exercise progression and should gradually increase time, distance and intensity of throwing sessions to avoid overload, with advice on the onset of fatigue to be used as an indicator to cease throwing and to minimise injury risk.

This is also the time for preseason screening to occur, with athletes deemed at risk, instructed in specific exercise programmes, to correct impairments, which are then closely monitored during the competitive season

8.7 Aspects in Rugby, Handball, Basketball, Volleyball and Tennis

8.7.1 Rugby

Maxillofacial and dental injuries occur in all contact sports as rugby, basketball, handball and volleyball. The most frequent causes of injury are a blow from another player. Because of this we should promote the use of mouthguards to sport participants.

Non-acute shoulder injury in rugby and American football, because of throwing, has a far lower incidence compared to baseball. This is of the weight of the ball resulting in lower velocity and a slower motion. This is decreasing the load on the soft tissue.

By video analyses of shoulder injury, it was found that direct impact to the shoulder, either through player-to-player contact or contact with the ground, is the main cause of acute shoulder injury. The main mechanisms of shoulder injury were contact with the ground with the shoulder/arm in horizontal adduction, flexion and internal rotation and impact to the lateral aspect of the shoulder with the elbow flexed and arm at the side. Methods to reduce injury risk, such as shoulder pads and tackle skills, require consideration.

Recent research by Bohu (2014) showed a 66 % shoulder dislocation recurrence rate in players with previous shoulder dislocation, in a five-season prospective study. For this reason they suggest extra attention from sports medicine professionals.

Rugby players are at higher risk in developing injury comparing to other team sports. Because of the high physical impact, there is also an increased chance for serious injury. Kew et al. (1991) concluded that 56 % of all rugby-related spinal cord injuries were preventable.

For the prevention of serious injury, the RugbySmart programme in New Zealand and BokSmart programme in South Africa are developed. This results mainly with the BokSmart programme in a decrease in serious injury in junior players.

The lower-extremity injuries are discussed in earlier chapters.

8.7.2 Handball

A recent study by Clarsen is following 206 elite male handball players throughout the season. The prevalence of shoulder problems was 28 % (95 % CI 25–31 %). The prevalence of substantial shoulder problems, defined as those leading to moderate or severe reductions in handball participation or performance, or to time loss, was 12 % (95 % CI 11–13 %). The problems seen are reduced glenohumeral rotation, external rotation weakness and scapular dyskinesis. Injury prevention programmes should incorporate interventions aimed at improving glenohumeral rotational range of motion, external rotation strength and scapular control as discussed in paragraph 7.3.

Concentric strength exercises for internal and external rotators on the nondominant side are suggested to prevent possible orthopaedic injuries associated with muscular bilateral asymmetry, together with functional exercises that improve eccentric rotation strength for the dominant side during rehabilitation and injury prevention programmes for handball athletes.

The lower-extremity injury is comparable with volleyball and basketball and is discussed below and in earlier chapters.

8.7.3 Volleyball

In volleyball, as well as in handball players, asymmetry was found in resting scapular posture. Clinicians should be aware that some degree of scapular asymmetry might be normal in some athletes. It should not be considered automatically as a pathological sign but rather an adaptation to sports practice and extensive use of the upper limb.

In some overhead activities such as volleyball and tennis, high elbow extensor to flexor ratio is seen.

Shoulder tendinitis secondary to the overhead activities of spiking and serving is also commonly seen and can be managed as described in earlier paragraphs.

Injury involving the distal branch of the suprascapular nerve, which innervates the infraspinatus muscle, has been increasingly described in volleyball players.

In the lower extremity, the jumper's knee is a frequent injury in volleyball and basketball players. Male genders, a high volume of volleyball training and frequency in match exposure were risk factors for developing jumper's knee. A review by Van der Worp suggests that horizontal landing poses are the greatest threat for developing patellar tendinopathy. A stiff movement pattern with a small post-touchdown range of motion and short landing time is associated with the onset of patellar tendinopathy. Learning to perform a flexible landing pattern seems to be an expedient strategy for reducing the risk for developing patellar tendinopathy. The findings in this review indicate that improving kinetic chain functioning, performing eccentric exercises and changing landing patterns are potential tools for preventive and/or therapeutic purposes.

The ACL rupture is also a very devastating injury that frequently occurs in most pivoting and jumping sports as rugby, handball, volleyball and basketball. The prevention of ACL injury is discussed in a previous chapter.

The most frequent injury in ankle sprain can be managed by stabilisation, proprioception and strength exercises and/or use of an ankle brace.

8.7.4 Tennis

Tennis is characterised by the use of a series of high-intensity and explosive actions, very short sprints, changes of direction and abrupt deceleration. This is resulting in a lot of physical stress.

The tennis serve is the most complex stroke in tennis. The complexity is because of the combination of core, limb and joint movement throughout the whole kinetic chain. The forces transfers from the lower body through the core into the upper body and then by the racket into the ball.

The serve motion is analysed in the same way as the throwing motion in baseball, shown in Fig. 8.1.

The differences in this motion consist in the plane of the motion, the contralateral arm tossing the ball, the extra lever of the tennis racket and the variety in speed, spin, direction and angle.

The three major types of serves used in tennis are the flat (limited spin), slice (sidespin) and topspin "kick" serves. The lower body does not show a real difference in the three serves.

The actual differences in the serve are seen in the upper part of the kinetic chain. It has to do with the angle of the racket changing forearm pronation and internal rotation of the shoulder.

A lot of research has been done on one- and two-handed backhands. This requires a different motor coordination. Two-handed backhand strokes rely more on trunk rotation for racket velocity generation, whereas one-handed backhand strokes rely more on segmental rotations of the upper limb. A tennis player using a two-handed backhand should learn early a slice one-handed backhand because of the different coordination patterns involved. Specific training exercises have to adjust this.

Frequent shoulder problems include internal impingement [30] and SLAP lesions as discussed in paragraphs 7.3 and 7.6

It has been shown that reduced grip forces decrease the vibration load on the arm and thus may prevent tennis elbow. Prevention programmes for the elbow in tennis players emphasise strength and endurance exercises for the entire upper extremity kinetic chain.

Tendinitis or luxation of the extensor carpi ulnaris (ECU) is a common injury in both the dominant and nondominant wrists due to the forearm and two-handed backhand. This is because during this stroke the wrist is in more ulnar deviation. Strengthening of the forearm flexors and extensors and modification of the technique can be a preventive measure in this injury.

As presented in recent research, functional asymmetry is thought to play a role in the risk for injury.

Balance training exercises seem to decrease lower-limb strength asymmetry in young tennis players.

Low back pain results from large loads in axial rotation and can result in shorting of hamstrings and limited hip rotation. Focus on core stability can prevent these problems.

Abdominal muscle strain is one of the most common injuries related specifically to tennis players because the abdominal musculature plays a significant role in the service motion. The trunk is extended and rotated to create energy for a powerful serve. Prevention as well as early rehabilitation starts with isometric as well with isotonic exercises.

The forehand stroke requires greater hip external rotation, which may increase anterior instability and posterior impingement.

A number of studies have been performed on the effect of heat stress in tennis players. In the situation when there is thermal discomfort, players decrease the pace of the match as a result of a drop in metabolic rate. Extra measures that players can make are the use of cold water, ice vests and towels, parasols and fans.

The change in equipment, for example, the use of low-compression balls in learning young children and adults, increases the time to return the ball and lengthens the rally time. This results in a positive overall experience and an improvement of technique.

The ITF (International Tennis Federation) has developed a complete website for tennis players and coaches. This includes a wide variety of practical information based on scientific research. The Tennis iCoach section (http://tennisicoach. com) contains technical, tactical, physical, mental and medical information for injury prevention. Also the Society for Tennis Medicine and Science (STMS) offers scientific-based performance education that gives all up-to-date knowledge.

8.7.5 Basketball

Most injuries in basketball are in the lower extremity, especially at the ankle and knee, and are discussed in earlier chapters in this book.

In jump-landing sports like basketball, volleyball and handball, the mechanism that results in injury is the jump-landing mechanism. Recent research generated more knowledge on this mechanism, and this is resulting in more preventive training programmes. Aerts et al. (2015) did a randomised controlled trail on 116 basketball athletes and found improvement after the completion of the prevention programme of valgus position and insufficient knee flexion and hip flexion. These factors are previously identified as possible risk factors for lower-extremity injuries.

The prevention of anterior knee pain is described in the paragraph on volleyball.

Ankle injuries are the most common high school basketball injury; there are prevention strategies for coaches such as prophylactic ankle bracing (PAB) and an ankle injury prevention exercise programme (AIEPP). In a survey by McGuine et al. (2013), among 480 basketball coaches, less than half of the coaches encouraged the use of PAB, and half did not utilise an AIEPP. Coaches had specific preferences for the type of AIEPP they would implement.

8.8 Passive Prevention

Potential risk factors for injury are physical conditioning, nutrition, hydration and playing environment [24], also discussed in the previous chapters. Young children often have less coordination and skill than older players. Strength and physical conditioning programmes for younger players should be aimed at developing general fitness and athleticism, with concentration on functional strength for pitching [24]. Nutrition and hydration are beneficial for general health as well as for recovery from pitching. Outdoor temperature and humidity are also factors for player health and risk of injury. When an athlete plays on several positions, the risk of injury is higher. The pitcher-catcher combination in particular has been suggested as a risk factor. If youth pitchers are bigger, faster and stronger because they developed early physically, they are usually overrepresented in key positions such as pitcher. Strong muscles with open growth plates may increase the risk of physeal plate injuries in a young pitcher's elbow or shoulder.

The injury in the adult and youth pitchers typically develops over a period of time; this makes early recognition the key to preventing permanent damage.

Conclusion
USA Baseball, Little League™ Baseball, Baseball Canada, ASMI, the American Orthopaedic Society for Sports Medicine and the American Academy of Pediatrics have published or implemented guidelines and rules for prevention of youth pitching injuries [24]. These guidelines focus on the awareness and prevention of overuse and consist of age-related recommendations concerning the number of pitches, the period of rest and the several throwing techniques. Unfortunately the recommendations are only available for baseball and difficult to extrapolate to other overhead sports. There is an urgent need to develop these guidelines for other (youth) overhead athletes.

References

1. Lyman S, Fleisig GS, Andrews JR, Osinski ED (2002) Effect of pitch type, pitch count, and pitching mechanics on risk of elbow and shoulder pain in youth baseball pitchers. Am J Sports Med 30: 463–468

2. Krajnik S, Fogarty KJ, Yard EE, Comstock RD (2010) Shoulder injuries in US high school baseball and softball athletes, 2005e2008. Pediatrics 125: 497–501

3. Register-Mihalik JK, Oyama S, Marshall SW, Mueller FO (2011) Pitching practices and self-reported injuries among youth baseball pitchers. A descriptive study. Athl Training Sports Health Care 4:11–20

4. Shanley E, Rauh MJ, Michener LA, Ellenbecker TS, Garrison JC, Thigpen CA (2011) Shoulder range of motion measures as risk factors for shoulder and elbow injuries in high school softball and baseball players. Am J Sports Med 39:1997–2006

5. Oyama S (2012) Baseball pitching kinematics, joint loads, and injury prevention. J Sport Health Sci 1:80–91

6. Kibler WB (1998) The role of the scapula in athletic shoulder function. Am J Sports Med 26:325–337

7. Van den Bekerom MP, Eygendaal D (2014) Posterior elbow problems in the overhead athlete. Sports Med Arthrosc 22(3):183–187

8. Eygendaal D, Safran MR (2006) Postero-medial elbow problems in the adult athlete. Br J Sports Med 40(5):430–434

9. Aguinaldo AL, Chambers H (2009) Correlation of throwing mechanics with elbow valgus load in adult baseball pitchers. Am J Sports Med 37:2043–2048

10. Lyman S, Fleisig GS (2005) Baseball injuries. In: Maffulli N, Caine DJ (eds) Epidemiology of pediatric sports injuries: team sports. Karger, Basel, pp 9–30

11. Huang YH, Wu TY, Learman KE, Tsai YS (2010) A comparison of throwing kinematics between youth baseball players with and without a history of medial elbow pain. Chin J Physiol 53:160

12. Fleisig GS, Chu Y, Weber A, Andrews JR (2009) Variability in baseball pitching biomechanics among various levels of competition. Sports Biomech 8:10–21

13. Myers Laudner KG, Pasquale MR, Bradley JP, Lephart SM (2006) Glenohumeral range of motion deficits and posterior shoulder tightness in throwers with pathologic internal impingement. Am J Sports Med 34:385–391

14. Hurd WJ, Kaplan KM, ElAttrache NS, Jobe FW, Morrey BF, Kaufman KR (2011) A profile of glenohumeral internal and external rotation motion in the uninjured high school baseball pitcher, part I: motion. J Athl Train 46(3):282–288

15. Reinold MM, Wilk KE, Macrina LC, Sheheane C, Dun S, Fleisig GS, Andrews JR (2008) Changes in shoulder and elbow passive range of motion after pitching in professional baseball players. Am J Sports Med 36(3):523–527

16. Sabick MB, Torry MR, Lawton RL, Hawkins RJ (2004) Valgus torque in youth baseball pitchers: a biomechanical study. J Shoulder Elbow Surg 13(3):349–355

17. Micheli LJ, Smith AD (1982) Sports injuries in children. Curr Probl Pediatr 12:1–54

18. Morrey BF, An KN (1983) Articular and ligamentous contributions to the stability of the elbow joint. Am J Sports Med 11:315–319

19. Tyler TF, Nicholas SJ, Lee SJ, Mullaney M, McHugh MP (2010) Correction of posterior shoulder tightness is associated with symptom resolution in patients with internal impingement. Am J Sports Med 38(1):114–119

20. Burkartt SS, Morgan CD, Kibler WB (2003) The disabled throwing shoulder: spectrum of pathology Part III: the SICK scapula, scapular dyskinesis, the kinetic chain, and rehabilitation. Arthroscopy 19(6):641–661

21. Cain EL Jr, Dugas JR, Wolf RS, Andrews JR (2003) Elbow injuries in throwing athletes: a current concepts review. Am J Sports Med 31:621–635

22. Gosheger G, Liem D, Ludwig K et al (2003) Injuries and overuse syndromes in golf. Am J Sports Med 31:438–443

23. Werner SL, Gill TJ, Murray TA, Cook TD, Hawkins RJ (2001) Relationships between throwing mechanics and shoulder distraction in professional baseball pitchers. Am J Sports Med 29(3):354–358

24. Fleisig GS, Andrews JR (2012) Prevention of elbow injuries in youth baseball pitchers. Sports Health 4(5):419–424

25. Fleisig GS, Kingsley DS, Loftice JW et al (2006) Kinetic comparison among the fastball, curveball, change-up, and slider in collegiate baseball pitchers. Am J Sports Med 34:423–430

26. Kibler WB, Chandler J (1995) Baseball and tennis. In: Griffin LY (ed) Rehabilitation of the injured knee. Mosby, St. Louis, pp 219–226

27. Hannon J, Garrison JC, Conway J (2014) Lower extremity balance is improved at the time of throwing in baseball players after an ulnar collateral ligament reconstruction when compared to pre-operative measurements. Int J Sports Phys Ther 9(3):356

28. Jaggi A, Lambert S (2010) Rehabilitation for shoulder instability. Br J Sports Med 44:333–340

29. Voight ML, Hardin JA, Blackburn TA, Tippett S, Canner GC (1996) The effects of muscle fatigue on and the relationship of arm dominance to shoulder proprioception. J Orthop Sports Phys Ther 23(6):348–352

30. Cools AM, Declercq G, Cagnie B, Cambier D, Witvrouw E (2008) Internal impingement in the tennis player: rehabilitation guidelines. Br J Sports Med 42(3):165–171

31. Cools AM, De Wilde L, Van Tongel A, Ceyssens C, Ryckewaert R, Cambier DC (2014) Measuring shoulder external and internal rotation strength and range of motion: comprehensive intra-rater and inter-rater reliability study of several testing protocols. J Shoulder Elbow Surg (American Shoulder and Elbow Surgeons [et al]) 23(10):1454–1461

Specific Aspects of Football in Recreational and Competitive Sport

9

Peter Angele, Helmut Hoffmann,
Andrew Williams, Mary Jones,
and Werner Krutsch

Key Points

1. In order to develop appropriate prevention and rehabilitation strategies it is important to understand the musculoskeletal adaptations found in football players which occur as a result of highly specific, stereotypical movement patterns.

2. These repetitive asymmetrical movement patterns result in musculoskeletal changes in the lumbar-pelvic- hip region and the lower limbs, with different adaptations found in the dominant kicking leg compared to the support leg.

3. Due to the stresses on the lumbar-pelvic- hip region as forces are transmitted to the lower limb core stability is essential to prevent compensatory movements which cause an increased injury risk.

4. Team prevention programmes, such as 11+ can lower injury incidence if conducted regularly, are of sufficient duration and include eccentric hamstring training, neuromuscular strategies regarding leg alignment and core stability work.

   Individualised prevention and treatment programmes are required to address individual deficits . When based on specific performance deficits and the individual's injury history these can help limit musculoskeletal problems and assist in the prevention of injuries and re-injuries.

P. Angele (✉)
Department of Trauma Surgery, FIFA Medical Centre of Excellence, University Medical Centre Regensburg, Franz-Josef-Strauss-Allee 11, Regensburg 93053, Germany

Sporthopaedicum Straubing/Regensburg, Hildegard-von-Bingen-Str. 1, Regensburg 93053, Germany
e-mail: dr-angele@t-online.de

H. Hoffmann
Eden Reha, FIFA Medical Centre of Excellence, Lessing-Str. 39-41, Donaustauf 93093, Germany

A. Williams • M. Jones
FIFA Medical Centre of Excellence,
Fortius Clinic, London,
17 Fitzhardinge Street, London W1H 6EQ, UK
e-mail: mary.jones@fortiusclinic.com

W. Krutsch
Department of Trauma Surgery, FIFA Medical Centre of Excellence, University Medical Centre Regensburg, Franz-Josef-Strauss-Allee 11, Regensburg 93053, Germany

© ESSKA 2016
H.O. Mayr, S. Zaffagnini (eds.), *Prevention of Injuries and Overuse in Sports: Directory for Physicians, Physiotherapists, Sport Scientists and Coaches*, DOI 10.1007/978-3-662-47706-9_9

9.1 Injury Incidence and Injury Pattern in Recreational and Professional Football

9.1.1 Introduction

Football is one of the most popular sports in the world. In 2006 the Federation Internationale de Football Association (FIFA) determined that over 265 million men and women of all ages across the world played football [17]. With a further five million officials and referees actively involved in the sport, it accounts for an involvement of nearly 4 % of the world's population. Although there are many health and well-being benefits to participation in regular sport, there are also risks. Despite it being viewed as a non-contact game, football is one of the main causes of sports injuries in all age groups. The injury rate for professional footballers is believed to be significantly greater than for workers in high-risk industrial occupations such as mining and construction [22].

9.1.2 Incidence of Injuries

The incidence of injuries in elite football has been variably reported as between 4.4 and 77 per 1000 h of playing time [3, 5, 12, 22, 28]. The variation in incidence is partly due to the method of reporting as some studies report all injuries whereas others only report injuries resulting in time loss from matches and/or training. The incidence also varies according to the playing level and is higher in professional football, particularly at international level [28].

3944 injuries were sustained during World Football Tournaments between 1998 and 2012, which involved a total of 1546 matches [28]. This equated to an injury rate of 2.6 injuries per match at a frequency of 77.3 injuries per 1000 playing hours although this figure dropped to 32.8/1000 (1.1 per match) if only those injures expected to prevent the player from playing or training were included.

The UEFA injury study has prospectively recorded match and training injuries resulting in time loss from training or matches from 26 professional clubs across 10 countries between 2001 and 2008. The overall incidence of injuries was eight per 1000 h, which on average meant two injuries per player per season [12]. These injury rates are comparable to those found in Sweden, Norway, and England.

More than 99 % of the world's football players are amateur, recreational, and youth players. Due to the variation in the provision of medical services to the wide range and distribution of these people, it is difficult to get an overall view of their injury rate. Figures from Spain support the view that injury rates are lower in recreational sports with a match injury incidence of 1.15/1000 h and an average of 0.11 injuries per player per year [24]. However, figures from Belgium suggest a 63 % higher injury risk for recreational players with 7.2 injuries per 100 players compared to 4.4 injuries in national players [5].

In the United States it is recognized that football is the third largest cause of high school sporting injuries, after wrestling and American football with an incidence of 2.39 per 1000 episodes of football participation [42]. In this study injury rates were similar between males and females, although a study in Denmark found time loss injury rates in adolescent females to be much higher at 9.7 per 1000 h [7].

9.1.3 Injury Pattern

The location of injuries sustained during differing levels of competitions is documented in Table 9.1. Regardless of age, gender, or level played, the majority of injuries sustained, on average over 70 %, are in the lower limb with the thigh, knee, and ankle most commonly affected [12, 23, 24, 28, 31, 42].

The second most frequently injured area is the head and neck with surveillance at the FIFA world cups showing injuries in this area to be significantly more frequent in women than men [28]. However, this gender difference was not apparent at adolescent level [42].

The majority of injuries sustained in elite football players are due to acute trauma with approximately 10 % due to overuse [12, 15]. Lesions occur most frequently in connective tissues (sprains, strains, dislocations, and other joint

Table 9.1 Percentage of injuries by injury site and competition type (in %)

	FIFA world cup male $n=543$ [28]	FIFA world cup female $n=220$ [28]	FIFA U-17 male $n=674$ [28]	FIFA U-17 female $n=223$ [28]	UEFA $n=4483$ [12]	England $n=6030$ [23]	Spanish amateur $n=243$ [24]	US high school boys 100 schools [42]	US high school girls 100 schools [42]
Head and neck	12	21	12	15	2.2	7[a]	9	13	15
Upper extremity	8	10	6	6	3.5	3	11	7	6
Trunk	7	7	7	9	7	2	7	5	4
Hip/groin	4	1	3	3	14	12	5	5	2
Thigh	20	10	13	7	23	23	10	14	12
Knee	13	13	11	13	18	17	31	15	22
Lower leg/TA	15	15	19	21	11	12	8	8	9
Ankle	14	15	22	17	14	17	12	22	25
Foot/toes	6	6	7	10	6	6	9	8	7

[a]Includes all spine injuries

injuries), vascular structures (hematomas, contusion, and lacerations), and bones. There is, in recent times, greater awareness of the need to identify and adequately treat concussion.

During World Football tournaments, contusions were the most frequent injury and accounted for approximately 50 % of injuries [28]. Whether this pattern is repeated at other levels is unclear as much of the literature only reports time loss injuries and therefore this would exclude many of these injuries. In amateur players in Spain, the incidence of contusions and hematoma is 23 % which, although it includes players who did not miss any training or matches, may still underreport the actual incidence as it is only recorded if medical attention was sought [24].

The incidence of fractures and dislocations is relatively low in elite football players and has been reported as between less than 1 and 6 % [3, 7]. Over a 15-year period in Japanese professional league players, 4.3 % of injuries were upper limb fracture/dislocations compared to 1.8 % in the lower limb [3]. In the UEFA study, 12 % of upper extremity injuries were shoulder dislocations, 25 % were upper limb fractures or bone stress injuries, and 5 % were AC joint dislocations [14]. Fifth metatarsal fractures accounted for 0.5 % of all injuries, and it has been estimated that a team can expect a fifth MT fracture every fifth season [15]. The overall rate of fracture in adolescents has not been reported, but in a study

of American high school football players, it found that 31 % of injuries resulting in time loss of more than 3 weeks were due to fractures [42].

A study investigating all football-related fractures in one area of Scotland over one season estimated that the annual incidence of football-related fractures was 0.7 per 1000 of the general population but the incidence per number of population playing football is not known. Of the 367 football-related fractures identified, 68 % were in the upper limb and 32 % in the lower limb. The most common fractures were finger phalanx (20 %), distal radius (20 %), and ankle (13 %) [35].

Muscle injuries are believed to account for over 30 % of time loss injuries in professional football [4, 13], 18 % in high school players [42], and 11 % in the female and 7 % in the male U-17 World Cup [28]. The incidence in amateur players is reported as 17 %, but again this could be artificially lowered by players not seeking medical treatment [24]. Ninety-two percent of muscle injuries affected the lower extremity and two thirds were acute injuries [4, 13]. The most commonly acutely affected muscles are the hamstrings (37 %), adductors (23 %), quadriceps (19 %), and calf (13 %). Respectively these accounted for 12 %, 14 %, 4 %, and 4 % of all injuries sustained. However, overuse injuries with a gradual onset were significantly more likely to affect the hip/groin [12]. Ninety-six percent of Achilles tendon disorders were classified as gradual onset tendinopathies with an incidence of 0.16/1000 h compared to 0.01/1000 for an Achilles rupture [34]. This equates roughly to one Achilles tendinopathy per team per season and a rupture per team every 17 seasons.

Ligament strains and ruptures have been reported as between 17 and 25 % of all injuries in professional football players. Ankle sprains (7 %) and knee MCL (5 %) are among the commonest injuries sustained [12]. Ligament sprains have been shown to account for over 67 % of all ankle injuries, of which over 80 % were to the lateral ligament complex [39]. A professional squad is likely to suffer four or five ankle sprains each season. Thirty-nine percent of all injuries in the knee are ligament sprains and three quarters of these involve the MCL. Acromioclavicular joint sprains accounted for 13 % of all upper limb injuries [14]. In Spanish male amateur football, ligament sprains and ruptures accounted for 32 % of all injuries with nearly 22 % of all injuries being knee ligaments, which is markedly higher than the incidence found in professional sport [24].

The ACL and the lateral ligament had the highest incidence of injury. Female football player has in general a significant higher risk for ACL injuries than male player [41]. Several other risk factors for ACL injuries are described in the literature like young age, match play [40, 41], and previous knee injuries [20]. Furthermore, rapid changes in proprioception and the increase in physical impact on a footballer's body result in higher injury rates [1, 33]. The participation of junior football players in professional senior football matches [16, 20] or implementation of a new professional league [31] results in short-term higher physical impact on player and can lead therefore to higher incidence of severe knee injuries. Therefore, injury prevention in professional football should be highlighted.

In high school adolescent football players, incomplete ligament sprains were the most frequent injury and accounted for 27 % of injuries [42]. In competitions girls were found to sustain complete knee ligament ruptures requiring surgery at a rate of 26.4/1000 h compared to 1.98/1000 h in boys.

9.2 Physical and Psychological Aspects in Football

Many investigations into the incidence of football injuries have attempted to understand which factors influence injury rates. The aspects that affect injury rates include the mechanism and timing of injury as well as player factors.

Injuries can occur at many different times and in different circumstances. The injury rate in matches is higher than training across all levels of competition with between 57 % and 63 % of injuries occurring during matches [5, 12, 23, 24]. In adolescents match injuries were three times more common than training injuries [42]. Upper limb injuries are seven times more likely to occur in matches and have a match incidence of 0.83/1000 h compared to a training incidence of 0.12/1000 h [14]. Muscle injuries in professional football players are six times more likely in match play (8.7/1000 h versus 1.4/1000 h) [4, 13], although in adolescents they were found to be more common in training [42].

Many studies categorize injuries into contact and non-contact injuries. The incidence of contact injuries varies considerably and has been reported as 25 % in the male amateur game [24] compared to 80 % at world tournaments [28]. Seventy percent of MCL and PCL injuries, 57 % of LCL, 37 % of ACL [32] and 58 % of ankle injuries [39] are contact injuries, whereas over 90 % of muscle injuries are non-contact [13].

Of the contact injuries reported, many occur as a result of foul play. At world tournaments the medical staff judged 47 % of the contact injuries to be caused by illegal play [28], which was similar to the level found in the Japanese league [3]. However, UEFA relied on the referee's judgment to determine if a foul had taken place and found that only 21 % of contact injuries have been caused this way [12]. Forty percent of ankle injuries were felt to be caused by fouls, but only 5.8 % of these were given red or yellow cards [39].

Several studies have investigated the timing of injuries, both within the games and in the longer term. There are increased injuries in training during the preseason period [23], and in particular Achilles tendon injuries are more common at this time [18]. Overall the injury frequency does not vary within each season, although there is believed to be a decreasing trend of injuries over several seasons [3, 28, 32, 39]. Several studies have found an increased frequency of injuries in the last 15 min of each half [3, 12, 23, 32]. Calf strains are most common in the last 15 min of the match [13], and there are significantly fewer ankle injuries in the first 15min of a match [39].

The rate of injuries can vary according to the position played. The prevalence of upper limb injuries in goalkeepers is between 10 and 25 % compared to 2–5 % in field players [14]. However, the overall injury rate for goalkeepers is 0.5-fold lower than that for field players and significantly lower for contact injuries and lower limb injuries [3].

It is believed that limb dominance can also affect the injury incidence rate with 50 % of injuries affecting their dominant side compared to 37 % in the non-dominant side [23]. 60 % of MCL injuries are in the dominant leg [32] as are 60 % of quadriceps injuries [13]. However, injuries to other muscle groups do not appear to be affected by leg dominance.

One factor that does increase injury rate is a previous injury. Overall injury recurrence rates have been reported as around 7–12 % [12, 23] in professional footballers but at less than 3 % in amateur players [24]. Upper limb injuries, knee MCL ligament injuries, and ankle sprains have been shown to have recurrence rates of 12 %, 11 %, and 10 %, respectively [14, 32, 39]. The recurrence rate for Achilles tendinopathy is 31 % if the recovery period is 10 days or less and 13 % if the recovery period is longer [18]. In the first year after concussion, there is a significant increased risk of sustaining a subsequent injury [34]. There is a well-recognized tendency to sustain injury soon after return to play after any recent injury not affecting the area of the new injury, presumably due to deconditioning, leading to the concept of a "second injury syndrome." Clearly delaying return to play until full recovery helps prevent this issue.

Another patient factor that can affect injury rate is age. In amateur players those aged 30 or over suffered 0.2–0.4/ 1000 h more injuries than those under 30 [24]. Muscle ruptures were higher in the older group but ligament injuries were lower. Professional players with Achilles tendinopathy were significantly older and had an average age of 27 years [18]. Newcomers in their first season of professional football were significantly younger (18.8 versus 26 years) than existing players and had lower training injuries, but there was no difference in overall match injury rates [30]. However, the younger group experienced less muscle and tendon injuries and significantly higher bone stress injuries.

The psychological aspects of injury prevention in football relate mainly to the abandonment of fair play rules and adoption of reckless play. On occasion we have all seen a player who seems to 'lose the plot' and behaves recklessly especially when the play involves tackling or occasional "off the ball" violence. There is no published evidence that the authors are aware of that shows a relationship between resting (non-playing) mental state and injury – either being injured or causing injury. The increasing desire to "win at all costs" mentality is obviously not helpful with regard to injury from foul play. Of course there is a role of football in promoting mental well-being.

9.3 Basic Aspects and Methodology of Prevention Conception in Football

9.3.1 Sport-Specific Adaptations on the Musculoskeletal and Myofascial System

Different sports and athletic disciplines – and specially ball team sports and football – are characterized by a multitude of highly specific, stereotypical patterns of movement. When the movements are performed at sufficient magnitudes for a long period of time, these sport-specific motor stimuli evoke specific responses in which certain biological structures undergo adaptations that enable the athlete to adequately "process" the loads. These changes affect bones, ligaments, and musculoskeletal and myofascial structures and are characterized in all sports by an asymmetrical distribution of loads between the right and left sides of the athlete's body, especially in football with normally dominant kicking and standing leg.

Generally the adaptations heighten the quality of the sport-specific movement patterns and thus have a positive effect on the athlete's performance in that particular sport. On the other hand, many of these adaptations cause changes in muscular loads and can sometimes lead to the overuse or unphysiologic loading of certain musculoskeletal structures. These loads may exceed the stress tolerance of the structures, resulting in muscular injury.

An awareness of sport-specific changes in the musculoskeletal system will make it easier for the members of the sports medicine staff, especially the physical therapists and rehabilitation trainers, to evaluate the structural and functional consequences of muscle injuries and formulate appropriate, complex treatment strategies. The following discussions are intended to alert coaches, team doctors, and therapists to the existence and importance of sport-specific adaptations and have a direct impact to the needs and the conception to prevention strategies.

9.3.2 Football-Specific Changes and Adaptations of the Musculoskeletal System

Taking football as an example of a sport with side-specific or asymmetrical stress patterns, we shall look at corresponding adaptation patterns that should be noted in the evaluation of injuries. The stereotypical loads that act on certain biological structures may vary greatly in football, both quantitatively and qualitatively, and reflect a long-term adaptation of the musculoskeletal system to a recurring stress pattern. Players with a symmetrical kicking technique tend to be the exception in this regard. Also, the play requirements and stereotypical movement patterns vary from one playing position to the next. This difference is particularly marked between the goalkeeper and field players but also exists among different playing positions on the field.

Sport-specific musculoskeletal adaptations are found in active football players as well as in players who retired from the sport years earlier. This is particularly important in physical therapy settings (provided by a doctor, therapist, or trainer) where it is important to consider whether the adaptive changes should be prophylactically "treated" and reversed, or at least limited, with the goal of preventing future degenerative problems. There is no generally valid recommended course of action, and management decisions should be made on a case-by-case basis depending on the extent of the changes and on individual physical factors.

9.3.3 Changes Caused by Contact of the Kicking Leg with the Ball

From a mechanical standpoint, the football player who kicks a ball is accelerating an approximately 350-g mass of a specified size and volume in a designated direction. This can be accomplished by various modes of ball contact, which impose corresponding mechanical loads on the striking area – the forehead for a header, the instep for a side-foot pass or shot, or the dorsum of the foot for shot "off the laces." The weight of the football, the air pressure in the ball, the contact time, and the speed change at ball contact are all mechanical variables that determine the nature of the mechanical loads acting on the muscles, bones, and joints. Changing any one of these variables will alter the mechanical stress configuration, producing positive or negative effects on musculoskeletal structures. Variation in these factors has hugely improved with modern ball manufacturing techniques.

Besides the magnitude of the mechanical stresses associated with ball contact, the number of repetitive stereotypical loads caused by ball contact within the physiologic range will also trigger degenerative and overload changes in the musculoskeletal system. When the foot strikes a football, the impact exerts a force of short duration (ball contact time of 11–15 ms, depending on ball pressure) that is opposite to the arched

Fig. 9.1 Intra-articular loads during a football instep kick [11]

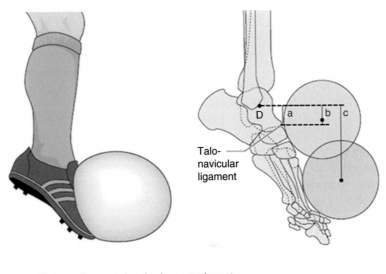

Talo-navicular ligament

a Distance from rotational axis to attachment of talonavicular ligament.

b Distance (lever length) to center of ball with correct kicking technique.

c Distance (lever length) to center of ball with incorrect kicking technique.

construction of the foot, giving rise to intra-articular shear forces. The mechanical reaction forces generated by the ball mass have a magnitude that is well within physiologic limits and generally do not exceed the stress tolerance of the biological structures. But when a large number of contacts are repeated over a long period of time, which may be measured in years, they create stimuli that act as repetitive microtrauma and will eventually evoke changes in the musculoskeletal system.

Naturally, the body tries to prepare for the sudden, brief tensile stresses caused by ball contact by strengthening the attachment sites of the talonavicular ligament. As the Sharpey fibers become stronger and more numerous, they create a mass effect that appears as a talar beak and/or tibial peak on X-ray films. This feature decreases the range of foot extension at the ankle joint. Moreover, kicking balls with a faulty, biologically unfavorable technique will quickly increase the tensile stresses on the talonavicular ligament to unphysiologically high levels that may exceed stress tolerance, resulting in an acute injury. Poor footwear can further exacerbate this adverse

change in the stability of the affected joint. While the soles of football shoes are designed to prevent excessive arch stresses in the extended foot (plantar flexion at the ankle joint), a poor kicking technique can still produce the adverse effects (Fig. 9.1). Striking the ball with the toe of the shoe as opposed to the laces has the effect of lengthening the lever arm of the ball-strike force and will multiply the torque and tensile stress acting on the talonavicular ligament, depending on the relationship of the extended lever arm to the lever-arm length of the talonavicular ligament. Under realistic conditions, the tensile stresses generated by executing a corner kick (initial ball speed of 50–80 km/h) are estimated at approximately 1200 N. This load is within physiologic limits and does not exceed the stress tolerance of the ligament. But a poor kicking posture will increase the tensile stress on the talonavicular ligament to as much as 3000 N, which approaches the stress limit and poses a risk of acute injury.

Besides direct changes to the ankle joint in response to kicking movements, football players also tend to develop asymmetrical muscle changes in the kicking leg and supporting leg. From a biomechanical standpoint, the kicking movement of the leg is an "open kinetic chain" action in which the foot is moved at maximum forward speed (moving point) while the hip is relatively stationary (fixed point). At the same time, every kicking movement will impose a "closed kinetic chain" type of load on the non-kicking side. In this case the foot is planted on the ground (fixed point) while the overlying structures of the pelvic-leg axis and torso are in motion (moving point) and must therefore be stabilized against gravity through complex coordination. Various neuromuscular control actions, especially those that stabilize the knee joint and the entire lumbar-pelvis-hip region, initiate long-term muscular adaptations to these football-specific movement patterns. Current research shows that active musculoskeletal structures progressively adapt to the characteristic movements that they perform and the loads associated with those movements to develop an optimized muscular response.

Reports in the literature describe significant muscular differences between the support leg and kicking leg in football players. Shooting the ball is a multiple-joint movement in which an (apparently) explosive extension of the knee is combined with active flexion of the hip and extension (plantar flexion) of the foot at the ankle joint. In the literature an increased maximum strength capacity and striking force of quadriceps muscle contraction during extension on the kicking-leg side are described, accompanied by an increased maximum strength and striking force of the knee flexors on the support side.

These general tendencies (quadriceps stronger on the kicking side, hamstrings stronger on the support side) vary in different playing positions according to the requirements of those positions. Goalkeepers show the greatest degree of extensor dominance. This results from their specialized ready position in which the knees are flexed almost 90°. For biomechanical reasons, this posture inhibits coactivation of the flexor muscles to stabilize the knee joints and eliminates the contribution of the hamstrings to knee extension. This is consistent with "Lombard's paradox," which states that the hamstrings contribute to knee extension only up to about 50–60° of knee flex-

ion. As a result, the goalie must rely entirely on the knee and hip extensors when initiating a jump from the ready position. This action requires great extensor power as a functional response to the football-specific demands of training and play. Strikers, on the other hand, must be able to react and accelerate their locomotor apparatus as quickly as possible. Hence, their flexor-extensor ratio allows for very effective coactivation of the flexor muscles and closely resembles the proportions that are found in sprinters.

Empirical observations on the degree of quadriceps muscle development in football players suggest additional neurophysiologic aspects and considerations relating to long-term functional adaptations. Although football players have greater quadriceps strength on the kicking side than on the support side, examination of the thighs in most players will show that the thigh circumference is slightly reduced in the area where the vastus medialis muscle is most fully developed. The variable "muscular configuration" of the quadriceps appears to be an adaptive response of the musculoskeletal system to years of locally varying, stereotypical functional demands (see above). Figure 9.2 shows the general lateralization of the quadriceps action that develops in the kicking leg of right-footed football players due to deficiency of the vastus medialis.

From a neurophysiologic standpoint, a "chronic" vastus medialis deficiency in the kicking leg can be explained by the reduced need for open kinetic chain activity to stabilize the knee joint in terms of tibial rotation relative to the femur in response to gravitational effects. The innervation pattern gradually adapts and becomes optimized for kicking a football. This process alters the relative contributions of the individual quadriceps muscles to the resultant quadriceps force. Deficiency of the vastus medialis tends to lateralize the quadriceps pull on the patella, thereby altering the kinematics of the femoropatellar joint (Fig. 9.2). If the lateralizing effect causes patellar rotation, this will reduce the area of contact between the retropatellar cartilage and femoral articular surface, and this could accelerate degenerative changes in the long term. By contrast, the loads on the supporting leg during

Fig. 9.2 Muscular configuration of the kicking-leg quadriceps [11]

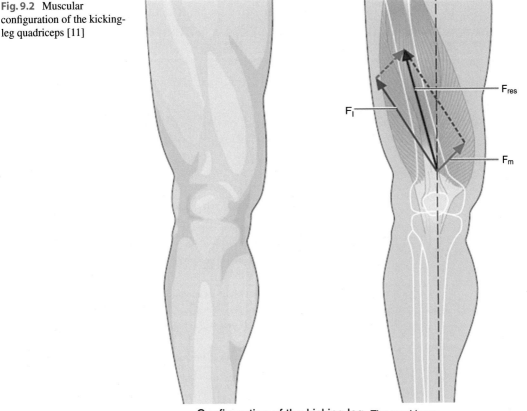

Configuration of the kicking leg. The quadriceps muscle and lateralization. Direction of quadriceps pull

F_{res} Resultant direction of quadriceps pull

F_l Direction of vastus lateralis pull

F_m Direction of vastus medialis pull

running and sprinting usually do not alter the physiologic pattern of intra-articular kinematics.

9.3.4 Support Leg Changes Caused by Kicking Technique

The changes in the kicking leg described above suggest that the contralateral support leg is subjected to different loads during the kicking of a football. It is interesting that all football players, regardless of performance level, tend to place their support leg in a very precise position when shooting the ball (i.e., when executing an instep kick or an inside/outside kick). This causes a highly consistent pattern of stereotypical loads

to act on the musculoskeletal structures (Fig. 9.3). To permit successful ball acceleration by the kicking leg with effective momentum transfer to the ball, the support leg must be planted next to the ball on the ground. The following observations are important in this regard:

• Football players plant their support leg next to the ball with remarkable consistency and precision. Tests have shown that interindividual differences from one ball contact to the next are less than 1 cm.
• Football players plant their support leg level with the ball (relative to the frontal plane).
• As the foot is planted on the ground, the body center of gravity shifts outward toward the

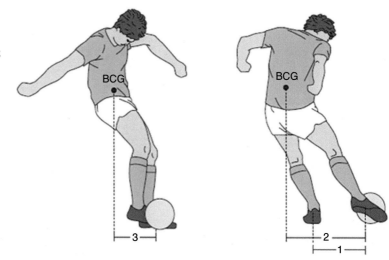

Instep kick viewed in the frontal and sagittal planes.
BCG, body's center of gravity.

1 Distance from the foot of the supporting
   leg to the center of the ball.
2 Distance from the BCG axis to the ball
   center of gravity.
3 Distance from the BCG axis to the foot
   of the supporting leg.

support leg, usually moving past the left knee or even farther laterally.

- The lateral distance of the support leg from the ball can vary markedly from one player to the next. Despite these differences, however, the individual movement patterns are carried out with great precision (intraindividual consistency). But the farther the support leg is placed from the ball, the greater the lateral shift of the body center of gravity. The joints along the left pelvic-leg axis must stabilize and compensate for this position and adapt to it over time.

These side-specific changes are most clearly appreciated in the ankle joint. The greater the lateralization of the pelvic-leg axis, the greater the lateral and shear forces acting on the joints of the foot. These forces will evoke long-term adaptations even in the absence of trauma or injuries. These changes are reflected not just in the stereotypical kicking actions that occur during training and play but also in ordinary walking and running. They document the overall adaptations of the pelvic-leg axis (Fig. 9.4).

9.3.5 Adaptations of the Lumbar-Pelvic-Hip Area

The foregoing neurophysiologic changes that occur on the supporting and kicking sides in response to playing stresses also induce long-term changes in the healthy lumbo-pelvic-hip region. The dominance of the powerful quadriceps and hip flexors (especially the iliopsoas muscle) on the kicking side causes the pelvis to tilt posteriorly on that side. This in turn causes an anterior pelvic tilt to develop on the opposite side in an effort to stabilize the body center of gravity. Often these changes are accompanied by a decreased range of motion in the sacroiliac joint on the kicking side. The asymmetrical range of motion, combined with the twisting of the hips, causes an apparent lengthening of the support leg axis and leads to functional pelvic obliquity. Additionally,

Fig. 9.4 Lateral shift of the support leg alignment in a football player during normal walking/running [11]

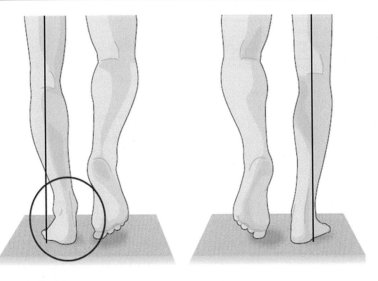

Lateral shift of the support-leg axis in a football player during gait.

Table 9.2 Adaptations of the kicking leg and supporting leg

Adaptations of the kicking leg	Adaptations of the support leg
Posterior pelvic tilt and inflare	Upright ilium and outflare
Decreased sacroiliac joint motion	Normal sacroiliac joint motion
Iliotibial tract	Increased valgus angulation at the knee
Vastus medialis atrophy	Groin problems
Patellar dyskinesia	Adductors (visceral causes)
Decreased range of plantar flexion	Increased external rotation of the foot
Supinator weakness or endurance loss due to eccentric muscle actions	Hyperpronation
Supinated limb position	Plantar insertional tendinopathy
	Pronated limb position

the new stress patterns are transmitted to the structures of the lumbar spine. As a result of the posterior pelvic tilt on the kicking side, physical examination of football players will often show that the lumbar spine is rotated to the right due to increased tension on the iliolumbar ligaments. The hip torsion may adversely affect various musculoskeletal structures (Table 9.2) and should be noted and evaluated by sports physiotherapists.

Note that the above listing covers expected gross musculoskeletal adaptations in football players. It does not provide a comprehensive differential diagnosis of sacroiliac, iliosacral, or other changes [10, 11].

9.4 Football-Specific Prevention Activities of Overload and Reducing the Risk of Injury in High-Performance Football

Different injury prevention strategies for professional football are available. In general, injury prevention includes all participants involved in the team sport (Fig. 9.5).

In the following section, this will be explained for "protective equipment." The industry has to provide the player with protective equipment, which is not only safe but also does not cause disturbance on the part of the player. The player has to wear sufficient protective equipment. Proper wearing of protective is standardized by FIFA and controlled by the referees. However, for training protective wear is not standardized; therefore, administrative authorities and medical team should work on improvement [29]. Training concepts and

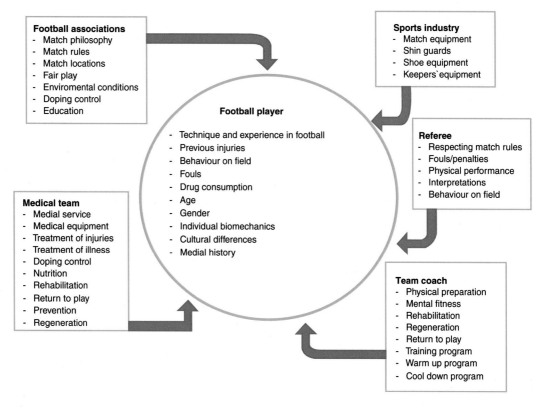

Football associations
- Match philosophy
- Match rules
- Match locations
- Fair play
- Enviromental conditions
- Doping control
- Education

Sports industry
- Match equipment
- Shin guards
- Shoe equipment
- Keepers`equipment

Football player
- Technique and experience in football
- Previous injuries
- Behaviour on field
- Fouls
- Drug consumption
- Age
- Gender
- Individual biomechanics
- Cultural differences
- Medial history

Referee
- Respecting match rules
- Fouls/penalties
- Physical performance
- Interpretations
- Behaviour on field

Medical team
- Medial service
- Medical equipment
- Treatment of injuries
- Treatment of illness
- Doping control
- Nutrition
- Rehabilitation
- Return to play
- Prevention
- Regeneration

Team coach
- Physical preparation
- Mental fitness
- Rehabilitation
- Regeneration
- Return to play
- Training program
- Warm up program
- Cool down program

Fig. 9.5 Injury prevention for football injuries [2]

warm-up programs represent an important purchase for injury prevention by team coaches or players themselves.

Several injury prevention programs in the form of training strategies [33] and warm-up exercises [37] are available in the literature. Particularly 11+ showed a huge reduction of injuries in football especially for severe knee injuries. These general injury prevention strategies should support team coaches to understand injury mechanism and to adapt training and warm-up exercises.

A complex concept for prevention, shown in Fig. 9.6, includes a variety of potential approaches that can certainly only completely be realized in a professional team environment with a professional infrastructure. The complex concept integrates interventions that can be scheduled in the normal training process of a football team (from teams of all performance levels) and interventions that have to be realized in addition to team training individually by a coach/rehab trainer or physiotherapist with the athletes. It has to be noted that the complex concept shows aspects

and dimensions, but does not claim completeness. The positive effects of the different interventions cannot be added up in their effects for injury prevention (Fig. 9.6).

9.4.1 Optimized Equipment

Biomechanical adaptations and changes in the musculoskeletal system of footballers are an important reason for side differences in leg alignment as described in 8.4. Depending on the severity of these adaptations, an individual adaptation of football shoes by orthopedic insoles is suitable to avoid unphysiologic mechanical stress on the musculoskeletal system and seems to be useful in a prevention perspective.

Given the anatomical predisposition and/or anatomically significant normal variants (e.g., side differences in foot length and width, volume variations due to Haglund exostosis, etc.), custom-made football boots are a suitable and necessary condition to avoid nonphysiological

Fig. 9.6 Dimensions of a complex intervention prevention strategy in football

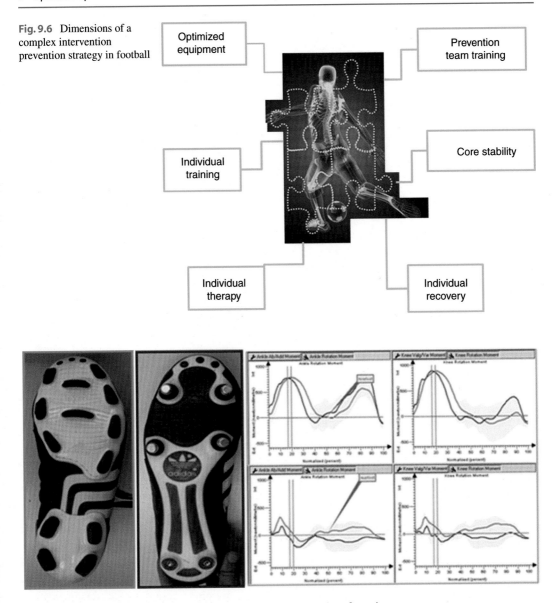

Fig. 9.7 Biomechanical parameters as a function of the different stud configurations

load dimensions. This service is currently available from most global manufacturers of football boots at least for the players with individual shoe contracts.

Not every shoe fits to every player. This should be tested under the biomechanical properties of different shoe styles/models in order to select a suitable shoe model for the individual athlete. The values of traction potential of the shoe model, which is defined by the type, number, and placement of studs on the shoe sole, should be included in this analysis (Fig. 9.7).

Football players have a long-standing tradition, due to the belief that foot sensitivity to the ball and hence ball control are enhanced, of wearing undersized boots – often two or even three sizes too small.

9.4.2   Prevention Team Training

This form of prevention training is the classic form of preventive interventions in football. In given and sufficiently documented evidence [33,

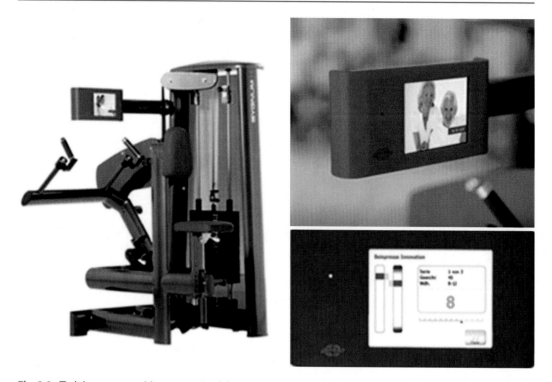

Fig. 9.8 Training systems with automated training management and documentation

37], positive impact on the incidence of injury can be detected, if:

- The corresponding exercises are conducted regularly over a long period of time.
- The exercises address appropriate dimensions of impacts (main topics are eccentric training of hamstring muscles, neuromuscular optimization of leg alignment of various landing stereotypes, and improvement of core stability).
- The duration of the prevention training allows an integration into the team training or as part of the warm-up program like FIFA 11+ (see Chap. 10).

9.4.3 Individualized Prevention Training

Prevention programs that can be integrated into the team training offer all players the same training stimulus; however, individual deficits of the players cannot be selectively altered or reduced.

Based on existing individual performance deficits and on the individual injury history, individual exercise programs can limit musculoskeletal and myofascial problems in mid and long term and thus also can contribute positively to injury prevention. In the professional infrastructure of top football clubs, selected locations with appropriate training equipment are available. Currently there is a tendency to establish networked training systems, thereby enabling a precise and individual training control and documentation (Fig. 9.8). These systems also allow the early detection of overtraining effects and can therefore contribute to prevention of injury in football.

9.4.4 Core Stability Training

Due to the described specific aspects of the lumbar-pelvic-hip region during football as a transition region of the trunk to the lower extremity, the core stability is of utmost importance. Deficits lead directly to compensatory movements and subsequently to unphysiological stress which has shown to exceed the stress tolerance of biological structures and therefore predisposes for increased injury. Good core stability thus constitutes the basis of the performance of the entire musculoskeletal system and correlates highly with the football-specific physical performance of the player. To limit and/or eliminate potential deficits in physical fitness of the individual players, evaluation methods such as FMS (Functional Movement Screening) [8] reveal core stability deficits (Figs. 9.9 and 9.10). An individualized exercise program directly tailored to these deficits can be developed (Fig. 9.11).

Deep Squad

Deep squad

Criteria	Point		Points	Classification
Heel lift-off grund	2		0	no deficits
Hip not in line with knee-joint	4		1–5	small deficits
Knees cannot be fixed in line with forefoot	3		6–14	Massive deficits, consult specialist (doctor, physio or athletic trainer) for creating individual trainings-program
Longbar cannot be fixed overhead (extended ellbows)	2			
Longbar touches door	3			**Cave:**
Longbar with lateral side movement	1			Pain during test: consult doctor

Figs. 9.9, 9.10 and 9.11 FMS and exercise consequences including documentation of the course (According to ZS Verlag/Dr. Kai-Uwe Nielsen)

Excercises after
deep squad
assessment

Fig. 9.9, 9.10 and 9.11   (continued)

9.4.5   Individualized Treatment Strategy

Especially in high level football with high training quantities/frequencies and enormous competition intensities, the endogenously reversible stress level of the musculoskeletal system exceeds the stress tolerance of the involved biological structures. As a consequence the joints of the lower extremities are heavily loaded in football training and matches. In order to achieve optimized regeneration/recovery of biological structures in professional football, the arthroligamentous balance has to be checked after training and competition. If necessary, the joint-specific arthroligamentous balance has to be reorganized with therapeutic "release" techniques (Fig. 9.12). This is essential to avoid pathologic changes in the intra-articular kinematics that could lead to increased risk of injuries.

Based on the individual injury history of the football player, regular assessment of potential consequences on the player's health should be performed. Whenever pathological changes can be detected, all variations of physical therapy, manual therapeutic techniques/interventions, and exercises of medical training therapy can be applied to counteract these changes.

9.4.6   Individualized Regeneration and Recovery

The current workout quantities in serious amateur and professional football do not allow more uncritical increase in training volumes. Performance improvements can be achieved by higher quality (not quantity) in training, which means more targeted training in the same time period. Alternatively, performance improvement can be realized by increased quantity in training. However, this can only be performed, when the regeneration and recovery after training loads can be shortened by various interventions. Only then the risk of "overtraining" with all negative consequences for the biologic structures can be reduced. A wide variety of regeneration interventions are currently offered by the industry;

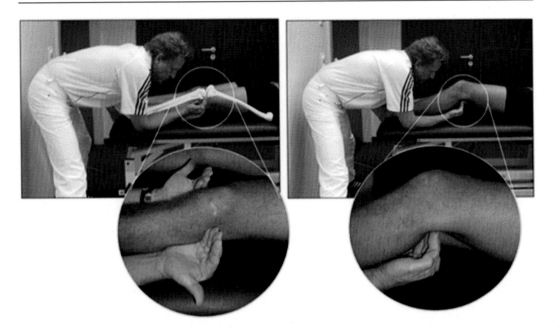

Fig. 9.12 Example of an arthroligamentous "release" technique for mobilization of the menisci (Allowance to use by Eden Reha)

Table 9.3 Regeneration and recovery interventions [21]

Stretching methods
Sleep
Local temperature applications
Hydration
Therapeutic interventions
Temperature management
Nutrition
Compression clothing
Water-immersion therapy

however, scientific proof of efficacy regarding the positive effects on regeneration/recovery after training/competition loads is not always proven. Following regeneration/recovery interventions have shown significant evidence on regeneration (Table 9.3).

Combination of different regeneration/recovery interventions will not necessarily show an additive effect in intensity and height. Often corresponding effects in combining different interventions can even eliminate their effects.

In this context, individual configurations of some interventions are conceivable and should be oriented on each individual athlete with appropriate psycho-vegetative positive attitude from the athlete to the methods/interventions preferred.

9.5   Different Prevention Activities for Male, Female, and Adolescent Athletes

Severe injuries like ACL ruptures have a multifactorial ethology [1]. It represents a combination of intrinsic and extrinsic risk factors with a specific injury mechanism.

"Age" represents one intrinsic risk factor. ACL ruptures are predominately detected in players between 15 and 25 years. One possible explanation could be the neuromuscular imbalance during length growth. Especially the transmission phase between youth football and adult football represents a risk period for ACL ruptures. Junior players, who played football with adults, showed an increased risk for ACL ruptures [2, 36]. This vulnerable time period for ACL injuries has to be respected in order to avoid injuries. Specific prevention programs for adolescent players should be implemented especially during training camps and in the preseason period, which predisposes the players to ACL injuries [31].

Table 9.4 Derivation of injury prevention exercises for ACL injuries in female football [26]

Injury mechanism	Underlying neuromuscular imbalance	Targeted Neuromuscular intervention component
Knee adduction during landing	"Ligament dominance"	Train for proper technique
Low flexion angle in landing	"Quadriceps dominance"	Strengthen posterior chain
Asymmetrical landings	"Leg dominance"	Train for side/side symmetry
Inability to control center of mass	"Trunk dominance" ("core dysfunction")	Core stability and perturb training

"Gender" represents another intrinsic risk factor. Different estrogen levels allow a strong influence on muscles and ligaments [19]. This predisposes female athletes to injuries to the musculoskeletal apparatus [9, 28, 38]. For females a two- to threefold increase in injury risk for ACL ruptures could be detected compared to males [41]. Especially the rate of non-contact ACL ruptures is significantly higher in female compared to male football. Male football players injure the ACL mainly on the kicking leg, female athletes predominately on the standing leg [6].

Hewett and colleagues [25, 26] recognized specific neuromuscular imbalances in female athletes and published specific injury prevention strategies for ACL injuries particularly adapted on injury mechanism and neuromuscular imbalance. Depending on injury mechanism and the causation of the imbalance in the knee during the injury, they defined four risk groups for ACL injuries (Table 9.4).

"Leg dominance" demonstrates the neuromuscular imbalance with valgus collapse during landing. The musculature cannot control the imbalance with high peaks of load to joints and ligaments. Training of landing technique after jump/turn around with avoidance of adduction moments in the knee joint represents the appropriate neuromuscular intervention training.

"Quadriceps dominance" represents the imbalance due to a higher quadriceps to hamstring ratio in female athletes. In addition, it was recognized that females show a landing maneuver in more extended knee position than males. Both lead to high anterior shift of the tibia with traction on the anterior cruciate ligament. Strengthening of the posterior chain (hamstring musculature) and training of landing maneuvers in more knee flexion are the appropriate neuromuscular intervention training.

The one leg dominance in females is significantly higher than in males ("leg dominance"). Muscle force, muscle reaction time, and muscle flexibility reveal a high side-to-side-difference in females, which predisposes for injuries. Training of leg symmetry represents the appropriate intervention training [27].

Athletes without adequate control of the trunk in 3D dimensions have a higher risk for injuries. One possible reason is a fast length growth during adolescence. This type represents the "core dominance" and can be influenced by core stabilization intervention training.

These specific ACL prevention exercises by Hewett and colleagues [25, 26] were also implemented in the complete warm-up program 11+ of F-MARC [37]. Particularly landing techniques after jumps or turn around, strengthening core stability, and strengthening the posterior chain in football player seem to be the key for the prevention of ACL injuries, especially for non-contact injuries. Together with training of leg symmetry starting in junior football levels and respecting match rules and fair play, prevention of ACL injuries may represent high effectiveness for all players.

References

1. Alentorn-Geli E, Myer GD, Silvers HJ, Samitier G, Romero D, Lazaro-Haro C, Cugat R (2009) Prevention of non-contact anterior cruciate ligament injuries in soccer players. Part 1: mechanism of injury and underlying risk factors. Knee Surg Sports Traumatol Arthrosc 17:705–729
2. Angele P, Eichhorn J, Hoffmann H, Krutsch W (2013) Prävention von vorderen Kreuzbandrupturen, SFA aktuell, Heft 26, Tuttlingen

3. Aoki H, O'Hata N, Kohno T, Morikawa T, Seki J (2012) A 15-year prospective epidemiological account of acute traumatic injuries during official professional soccer league matches in Japan. Am J Sports Med 40:1006–1014

4. Bengtsson H, Ekstrand J, Hägglund M (2013) Muscle injury rates in professional football increase with fixture congestion: an 11-year follow-up of the UEFA Champions League injury study. Br J Sports Med 47:743–747

5. Bollars P, Claes S, Vanlommel L, Van Crombrugge K, Corten K, Bellemans J (2014) The effectiveness of preventive programs in decreasing the risk of soccer injuries in Belgium: national trends over a decade. Am J Sports Med 42:577–582

6. Brophy R, Silvers HJ, Gonzales T, Mandelbaum BR (2010) Gender influences: the role of leg dominance in ACL injury among soccer players. Br J Sports Med 44:694–697

7. Clausen MB, Zebis MK, Møller M, Krustrup P, Hölmich P, Wedderkopp N, Andersen LL, Christensen KB, Thorborg K (2014) High injury incidence in adolescent female soccer. Am J Sports Med 42:2487–2494

8. Cook G (2003) Athletic body in balance. Human Kinetics, Champaign

9. Dvorak J, Junge A, Grimm K (2009) F- Marc-Football medicine manual, 2nd edn. RVA Druck und Medien AG, Altstätten

10. Eder K, Hoffmann H (2006) Verletzungen im Fußball – vermeiden – behandeln – therapieren. Elsevier Verlag, München

11. Eder K, Hoffmann H (2010) Physikalische und physiotherapeutische Maßnahmen und Rehabilitation. In: Müller-Wohlfahrt H-W, Überacker P, Hensel L (eds) Muskelverletzungen im Sport. Georg Thieme Verlag, Stuttgart/New York, pp 313–321

12. Ekstrand J, Hägglund M, Waldén M (2011) Injury incidence and injury patterns in professional football: the UEFA injury study. Br J Sports Med 45:553–558

13. Ekstrand J, Hägglund M, Waldén M (2011) Epidemiology of muscle injuries in professional football (soccer). Am J Sports Med 39:1226–1232

14. Ekstrand J, Hägglund M, Törnqvist H, Kristenson K, Bengtsson H, Magnusson H, Waldén M (2013) Upper extremity injuries in male elite football players. Knee Surg Sports Traumatol Arthrosc 21:1626–1632

15. Ekstrand J, van Dijk CN (2013) Fifth metatarsal fractures among male professional footballers: a potential career-ending disease. Br J Sports Med 47:754–758

16. Engebretsen AH, Myklebust G, Holme I, Engebretsen L, Bahr R (2011) Intrinsic risk factors for acute knee injuries among male football players: a prospective cohort study. Scand J Med Sci Sports 21:645–652

17. Federation Internationale de Football Association. Big count: FIFA survey. Accessed on 1/12/2014 at http://www.fifa.com/worldfootball/bigcount/index.html

18. Gajhede-Knudsen M, Ekstrand J, Magnusson H, Maffulli N (2013) Recurrence of Achilles tendon injuries in elite male football players is more common

after early return to play: an 11-year follow-up of the UEFA Champions League injury study. Br J Sports Med 47:763–768

19. Grimm K (2007) Fragen und Antworten zum Frauenfußball. In: Grimm K, Kirkendall D (eds) Gesundheit und Fitness für Frauenfußballerinnen – Ein Leitfaden für Spielerinnen und Trainer. Rva Druck und Medien AG, Altstätten

20. Hägglund M, Walden M, Ekstrand J (2006) Previous injury as a risk factor for injury in elite football: a prospective study over two consecutive seasons. Br J Sports Med 40:767–772

21. Hausswirth C, Mujika I (2013) Recovery for performance in sport. Human Kinetics, Champaign

22. Hawkins RD, Fuller CW (1999) A prospective epidemiological study of injuries in four English professional football clubs. Br J Sports Med 33:196–203

23. Hawkins RD, Hulse MA, Wilkinson C, Hodson A, Gibson M (2001) The association football medical research programme: an audit of injuries in professional football. Br J Sports Med 35:43–47

24. Herrero H, Salinero JJ, Del Coso J (2014) Injuries among Spanish male amateur soccer players: a retrospective population study. Am J Sports Med 42:78–85

25. Hewett TE, Di Stasi S, Myer GD (2013) Current concepts for injury prevention in athletes after anterior cruciate ligament reconstruction. Am J Sports Med 41:216–221

26. Hewett TE, Ford KR, Hoogenboom BJ, Myer GD (2010) Understanding and preventing ACL injuries: current biomechanical and epidemiologic considerations – update 2010. N Am J Sports Phys Ther 5:234–251

27. Hewett TE, Myer GD, Ford KR et al (2005) Biomechanical measures of neuromuscular control and valgus loading of the knee predict anterior cruciate ligament injury risk in female athletes: a prospective study. Am J Sports Med 33:492–501

28. Junge A, Dvorak J (2013) Injury surveillance in the World Football Tournaments 1998–2012. Br J Sports Med 47:782–788

29. Klügl M, Shrier I, McBain K, Shultz R, Meeuwisse WH, Garza D, Matheson GO (2010) The prevention of sport injury: an analysis of 12,000 published manuscripts. Clin J Sport Med 20:407–412

30. Kristenson K, Waldén M, Ekstrand J, Hägglund M (2013) Lower injury rates for newcomers to professional soccer: a prospective cohort study over 9 consecutive seasons. Am J Sports Med 41:1419–1425

31. Krutsch W, Zeman F, Zellner J, Pfeifer C, Nerlich M, Angele P (2014) Increase in ACL and PCL injuries after implementation of a new professional football league. Knee Surg Sports Traumatol Arthrosc. Epub ahead of print

32. Lundblad M, Waldén M, Magnusso H, Karlsson J, Ekstrand J (2013) The UEFA injury study: 11-year data concerning 346 MCL injuries and time to return to play. Br J Sports Med 47:759–762

33. Mandelbaum BR, Silvers HJ, Watnabe DS, Knarr JF, Thomas SD, Griffin LY, Kirkendall DT, Garrett W Jr (2005) Effectiveness of a neuromuscular and

proprioceptive training program in preventing anterior cruciate ligament injuries in female athletes: 2-year follow-up. Am J Sports Med 33: 1003–1010

34. Nordström A, Nordström P, Ekstrand J (2014) Sports-related concussion increases the risk of subsequent injury by about 50% in elite male football players. Br J Sports Med 48:1447–1450

35. Robertson GA, Wood AM, Bakker-Dyos J, Aitken SA, Keenan AC, Court-Brown CM (2012) The epidemiology, morbidity, and outcome of soccer-related fractures in a standard population. Am J Sports Med 40:1851–1857

36. Söderman K, Pietilä T, Alfredson H et al (2002) Anterior cruciate ligament injuries in young female playing soccer at senior levels. Scand J Med Sci Sports 12:65–68

37. Soligard T, Myklebust G, Steffen K et al (2008) Comprehensive warm-up programme to prevent injuries in young female footballers: cluster randomised controlled trial. BMJ 337:a2469. doi:10.1136/bmj.a2469

38. Tegnander A, Olsen OE, Moholdt TT, Engebretsen L, Bahr R (2008) Injuries in Norwegian female elite soccer: a prospective one-season cohort study. Knee Surg Sports Traumatol Arthrosc 16:194–198

39. Walden M, Hägglund M, Ekstrand J (2013) Time-trends and circumstances surrounding ankle injuries in men's professional football: an 11-year follow-up of the UEFA Champions League injury study. Br J Sports Med 47:748–753

40. Walden M, Hägglund M, Magnusson H, Ekstrand J (2011) Anterior cruciate ligament injury in elite football: a prospective three-cohort study. Knee Surg Sports Traumatol Arthrosc 19:11–19

41. Walden M, Hägglund M, Werner J, Ekstrand J (2011) The epidemiology of anterior cruciate ligament injury in football (soccer): a review of the literature from gender-related perspective. Knee Surg Sports Traumatol Arthrosc 19:3–10

42. Yard EE, Schroeder MJ, Fields SK, Collins CL, Comstock RD (2008) The epidemiology of United States high school soccer injuries, 2005–2007. Am J Sports Med 36:1930–1937

Specific Aspects of Alpine Skiing in Recreational and Competitive Sport (FIS)

10

Hermann O. Mayr, Martin Auracher, Max Merkel, Florian Müller, and Karlheinz Waibel

---

Key Points

1. To develop efficient prevention strategies of injuries and overuse in alpine skiing, one must know the underlying mechanisms as thoroughly as possible.
2. In alpine skiing speed and turn radius determine the resulting forces which must mainly be taken by the legs and torso. Thus, high demands for the skier's complex strength abilities arise.
3. An essential prerequisite for the practice of alpine skiing at the lowest possible risk is general endurance and strength training.
4. Proper muscle function is based upon its support and sheath function as well as the gliding capacity of the surrounding tissue layers.
5. To prevent overload damage in competitive sports, it is necessary to accelerate regenerative processes using therapeutic measures.

H.O. Mayr (✉)
Department of Orthopaedics and Traumatology,
University Hospital Freiburg,
Hugstetter Straße 55, Freiburg 79106, Germany

Schoen Clinic Munich Harlaching, Munich, Germany
e-mail: hermann.mayr@uniklinik-freiburg.de

M. Auracher • M. Merkel
Rehabilitation Center, Osteo Zentrum Schliersee,
Perfallstraße 4, Schliersee 83727, Germany
e-mail: martin.auracher@osteo-zentrum.de;
max.merkel@osteo-zentrum.de

F. Müller
Physiotherapeutic Practice, Alternative Therapy,
Kranzerstr. 13, Bayrischzell 83735, Germany
e-mail: fm@praxis-mueller-bayrischzell.de

K. Waibel
German Ski Federation, Science and Technology,
Hubertusstrasse 1, Planegg 82152, Germany
e-mail: Charly.waibel@ski-online.de

10.1 Injury Incidence and Injury Patterns in Recreational and Professional Skiing

10.1.1 Injury Incidence in Recreational Skiing

Alpine skiing in its numerous manifestations and variations keeps fascinating a large number of people of age and skill groups (Figs. 10.1, 10.2 and 10.3). Being active outdoors and indulging in surrounding nature, though, apart from pleasure, also carries risk of injury and excessive strain.

In Germany accidents in skiing have been documented and analyzed by an "Evaluation Authority for Skiing Accidents" (ASU) since 1979.

The first registration season has since been used as basis for the illustration of the developments in injury occurrence in ski sports. Despite a slight increase in the past season 2012/2013,

© ESSKA 2016
H.O. Mayr, S. Zaffagnini (eds.), *Prevention of Injuries and Overuse in Sports: Directory for Physicians, Physiotherapists, Sport Scientists and Coaches*, DOI 10.1007/978-3-662-47706-9_10

the amount of injured skiers per 1.000 has decreased by more than 58 % since 1979/1980.

In spite of the overall positive developments, injuries in alpine skiing are still a challenge for the people and institutions wanting to contribute to injury risk minimization. Because ski sports can only fully unfold their positive effects with further reducing injury risk and injury severity. In

Fig. 10.1 Alpine slope skiing

the 2012/2013 season, the "Evaluation Authority for Skiing Accidents" registered a slight increase of injured skiers. Projected onto the overall collective of approximately 4,2 million German skiers, between 41.000 and 43.000 skiers can be assumed to have sustained an injury that required medical attention [33].

When the total number of injuries is subdivided by involved body regions, the knee is still the most frequently affected joint, distantly followed by the shoulder (Fig. 10.4). Subdivided by sex, the dominance of knee injuries in women is prominent with close to 50 %. Even so, in men they are still close to 30 %. Considering the development of injury distribution to the various body regions, a slight decrease of knee injuries with a simultaneous increase of shoulder and arm, torso, and hip injuries becomes apparent. The reasons therefore have not been finally resolved since larger fluctuations can be recurrently observed. Yet, there is evidence that the course of collision accidents may be a reason for these developments.

10.1.2 Injury Incidence in Ski Racing

The high injury risk in competitive skiing also poses a serious problem for the athletes. Since 2006, the FIS, together with the Oslo Sports

Fig. 10.2 Alpine free ride skiing

Fig. 10.3 Downhill ski racing

Fig. 10.4 Distribution of ski injuries on body parts

Body part injured

Trauma Research Center (OSTROC), has been conducting an investigation to examine injury issues in the different Ski World Cup disciplines, to determine risk factors, and to develop efficient countermeasures.

The summary of injury frequency clearly shows the large count of injuries with at least 1–3 days of training and competing pause and severe injuries. Severe injuries are defined by a resulting training and competing pause of more than 28 days (Fig. 10.5).

The data demonstrate that almost one in every three athletes sustains an injury every season.

Considering injuries divided by gender, a slightly higher risk for injury among men is apparent. Altogether the development is subject to large fluctuations, thus making evaluation or cause analysis very hard (Fig. 10.6).

Fig. 10.5 Reported injuries per 100 athletes in alpine ski racing

Fig. 10.6 Reported injuries male vs. female in alpine ski racing

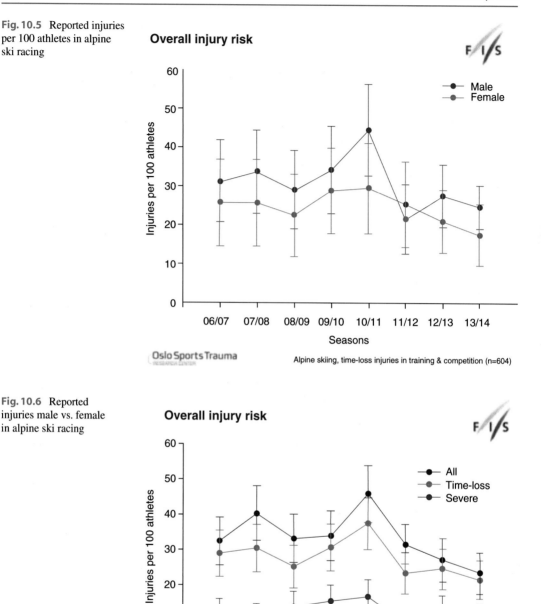

Examined by injury localization, the data show the dominant role of knee injuries, analogous to recreational skiing. The data analysis yields a 39 % ratio of knee injuries. The ratio even rises to 60 % regarding severe injuries only.

According to the data the anterior cruciate ligament is involved in nearly half of all severe injuries and thus becomes the focus of all deliberations concerning underlying causes and injury mechanisms as well as appropriate prevention strategies.

10.1.3 Injury Patterns in Recreational and Professional Skiing

To develop efficient prevention strategies, one must know the underlying mechanisms as thoroughly as possible. Due to the dominant role of the injuries to the anterior cruciate ligament, the most work exists on this topic.

A common cause for injuries in recreational skiing is a sudden change of direction of the lower limbs in relation to the torso. In the flexed knee they result in external rotation of the lower leg with concomitant opening up of the joint on the inner side ("valgus stress"), so that the otherwise stable ligament can eventually no longer withstand the present forces. Internal rotations with an outward tilt ("varus stress") are also possible. High velocity is not necessary here – this mechanism often leads to anterior cruciate ligament injuries at minor speed or even while standing.

The phantom foot is claimed to be the most common mechanism for ACL injuries in recreational skiing. In this situation, the skier is out of balance backward with the hips below the knees. The uphill arm is back, and the upper body generally faces the downhill ski. The injury occurs when the inside edge of the downhill ski tail engages the snow surface, forcing the knee into internal rotation in a deeply flexed position. The ski acts as a lever to twist or bend the knee, hence the term "phantom foot" [24].

A comprehensive analysis of injury mechanisms in ski racing was conducted by the Bahr group in cooperation with the FIS [5]. In this case an international group of seven experts, all biomechanists from the field of ski sports or sports physicians, examined twenty different cases of anterior cruciate ligament injuries that had been recorded within the FIS Injury Surveillance Study (ISS). Furthermore, all cases had been video documented. The experts analyzed the videos and described the injury mechanism concerning the current ski situation, behavior of the skier, and biomechanical conditions.

The group concluded that the majority of injuries can be described with three respective mechanisms. The most common one was the so-called slip-catch where the outer ski catches the inside edge, forcing the outer knee into internal rotation and valgus (Fig. 10.7). A similar loading pattern was observed for the dynamic snowplow. Injury prevention efforts should focus on the slip-catch mechanism and the dynamic snowplow [8].

The third described mechanism is typical of jumps. The athlete lands backward weighted after the jump with load to the ski tail at landing, giving it a forward rotation. The athlete tries to bring the body's center of gravity frontward again to regain balance. The strain to the anterior cruciate ligament develops from the combination of tibiofemoral compression, the "boot-induced anterior drawer" and the quadriceps anterior drawer.

10.2 Physical and Psychological Aspects in Recreational and Professional Skiing

A resilient immune system is an important factor for injury prevention. In this context stands a balanced diet and a healthy lifestyle in general (sleep, stress, etc.) [36]. The natural cold environment may increase the risk for hypothermia as an athlete does not have a sense of thirst although he is sweating due to the increased activity. Another typical characteristic of the cold air at higher altitude is the lack of humidity. The dry air needs to be moistened with every breath the athlete takes. This causes another source for loss of humidity that needs to be considered [18]. Sufficient fluid intake, especially at high elevations, is recommended.

Alpine skiing technique has changed considerably since the 1990s. Riding tapered skis, the so-called carving technique, has become more and more popular since then and has widely superseded skiing with non-tapered skis to present. The carving technique makes it possible for skiers of nearly all skill levels to experience the appeal of radial acceleration. The power necessary to initiate and modulate the turn with earlier techniques is largely omitted today. Albeit the skier is exposed to high forces resulting from tight curve radius and small sliding portion with extensive traction rise. Speed is a substantial risk factor concerning the carving technique even more than the technique without tapered skis.

Fig. 10.7 Slip-catch (right knee). (**a**) (2400 ms), the skier is out of balance backward and inward in the steering phase out of the fall line. (**b**) (2120 ms), as the skier tries to regain snow contact with the unweighted outer ski (**c**), he extends his right knee (**d**)

The cornering forces increase exponentially with speed. A study at the University of Salzburg, commissioned by the FIS, could provide evidence for this association in high-performance competitive skiing, also [32].

Modern slope preparation enables comfortable skiing and entices to high speed. This risk factor is concurrently the main fascination of slope skiing. Apart from the sidecut, the width and preload are of essential physical importance. Wide skis enable a stronger tilt and thereby a larger edging angle.

Due to the high rate of injuries in competitive alpine skiing, the FIS regulations for ski geometry were adapted following the present investigations of the University of Salzburg [15]. Present-day competition skis are less tapered and are altogether narrower. In recreational skiing so-called "Rocker"-skis are increasingly gaining acceptance. These are wide skis that are slightly bent upward in the front ski and thus have features similar to carving skis, even without greater sidecut. "Rocker"-skis afford pleasure off the slopes, also, and facilitate skiing in deep powder snow through more float. The "Rocker"-ski combines the advantages of the carving ski with the advantages of the conventional ski. Particularly the risk to catch an edge is substantially lower with a "Rocker"-ski compared to a carving ski. Slipping away of a ski followed by uncontrolled edge grip, known under the term "Slip Catch" in ski racing, is also a common injury mechanism in recreational skiing on distinctively tapered skis.

The work of Babiel et al. [3] can give an impression on the resulting forces. These investigations could determine that in competitive ski racing the leg is loaded with compression forces of up to 7000 N during turns. This load is mainly carried by the outside leg.

Depending on the level of difficulty of slope, race course, or mountain, all skiers are exposed to the same physical demands. Yet the athletes' constitution and stamina differ.

The physical requirements for alpine skiing and the skier's individual abilities must be brought in line. Precautious self-assessment and assessment of these facts is the first step to reasonable prevention. Speed and turn radius determine the resulting forces which must mainly be taken by the legs and torso. Thus, high demands for the skier's complex strength abilities arise. Apart from sufficient sense of movement, the athlete needs a capable musculoskeletal system. Compression and shear forces and bending and rotation moments are particularly applied to lower extremity joints and bony structures as well as the spine [39]. These loads can only be partially compensated by muscles and tendons and thus require a resilient passive locomotor system. Decent aerobic endurance abilities are an essential prerequisite for optimal injury prevention and protection from overload damage in both recreational and, especially, competitive alpine skiing [44].

Powerful and fully functional muscles are to be obtained and preserved by appropriate training and immediate preparative measures, such as warm-up exercises. The forced posture of the foot in the ski boot and the boot's notch effect at the edge lead to exceptional strain in alpine skiing. Permanent flexed knee position results in large tensile forces to the patellar tendon. Extensive forces and moments are put onto the knee through eccentric loading. The spine is exposed to great compression and shear forces due to the neutralization of its physiological S form while skiing in hunched posture. The upper extremity and head are mainly at risk in the case of fall or collision.

Next to the physical condition, the mental condition is crucial to injury and overload prevention.

The balance between fascination and risk has a key function. In recreational skiing the fascination mainly results from the experience of nature, motion dynamics, and speed.

The fascination in competitive skiing results from the desire to win and the will "to push the limits." Competitive skiers show an over proportional risk disposition. The successful ski racer knows how to manage risk on the borderline. Individual goals are essential criteria for prophylaxis in recreational and competitive skiing. Fatigue and exhaustion are accompanied by reduced concentration as well as coordination and are thus substantial risk factors. Prevention already begins with appropriate choice of equipment. Visual aspects and vanity should be set behind functional aspects. In back country skiing we should not only be guided by our emotions but also by the current avalanche warning, the weather conditions, and the terrain and snow composition.

10.3   General Aspects in Training and Competition Considering Various Techniques and Disciplines of Alpine Skiing

10.3.1   General Technical Training Aspects of Alpine Skiing

An essential prerequisite for the practice of alpine skiing at the lowest possible risk is general endurance and strength training. Warm-up of the musculoskeletal system is necessary before physical exercise and after longer pauses. Sport specific training is intended to achieve the desired level of performance. Further, the anticipation of hazardous situations is to be learned. The goal for technical sports training is executing the basic situative challenges instinctively, without activation of higher levels of awareness. Automation of motion sequences is the keyword. The capability to meet these demands with this method consists of repeating the supposed intended movement often enough to be able to execute it automatically under changing conditions. This approach is based on the principles of instruction and repetition. The exercise instructor gives motion tasks and corrects their execution. This instructor feedback does not require integral neuromotor processes; thus, their formation is not furthered. This model of learning is commonly found in recreational skiing and classic skiing instruction. In modern motion studies, elements of differential learning are used [31]. The trainer creates situations for the athlete and helps him find solutions. This form of intuitive learning focuses on experiencing and mastering the individual motion optimum by executing the movement in various, even extreme forms. Errors are not avoided; instead they help develop individual motion strategies. A training method is serving to form neuromotor pathways and, thus, multiple neuromotor connections for automated handling of as many situations as possible. This stimulates a self-organizing learning process, eventually making learning by sole motion execution possible, even without a trainer [40]. Alpine skiing is essentially a technomotoric sport. Learning of sports-specific motion sequences is also an elementary basis of this sport concerning prevention [37]. This form of training can be effective in recreational skiing, too. Recreational skiers are also confronted with different challenges, such as different terrains, slope conditions, snow quality, and slope frequenting by other skiers. For the purpose of optimal injury and strain prevention, a situationally adapted technique is helpful. The athlete is to learn and consider signs of fatigue within the scope of risk management.

10.3.2 General Aspects in Training and Competition in Various Disciplines of Alpine Skiing

According to Bahr et al. [5], the risk factors for a sports injury are derived from the intrinsic risk factors, which are specific to the athlete, and the extrinsic risk factors, depending on equipment and conditions. By means of a triggering event, a combination of these factors leads to an injury. A similar coherence can be assumed for overload syndromes.

Endurance capacity is also essential in terms of prevention. The competitive alpine skier's metabolism not only has to be able to mobilize enough energy to meet the specific demands but must also provide for a muscular milieu, con-

cerning pH regulation, that can enable required fine motor skills [41].

Which active mechanisms serve the prevention of injuries and excessive strain with due regard to the different competitive disciplines?

Slalom: particular requirement for this discipline is strength, explicitly springiness, since high-frequency movement cycles take place with turn durations of less than a second. The main challenge for the slalom skier is maintaining dynamic balance under ever changing conditions and the demand to lose as little speed as possible. In order to do so, he must constantly adapt the body's center of gravity and, thus, the point of load to the ski in the frontal and sagittal plane to the specific situation. These measures are used to precisely keep the ski on the edge to minimize friction and to utilize the ski's dynamic potential. The competitions last about 50 s. In this context the following factors of active prevention are essential:

- Optimization of mental and motor agility using training methods from the fields of coordination training or Life Kinetik® [17]
- Training of springiness under consideration of high contraction speeds (no terminal extension of joints, no maximal motion speed)
- Optimization of the interaction between eccentric and concentric motion sequences
- Specific training of endurance and force perseverance
- Torso-stabilizing training to counteract the whiplash-like inertia torque during turn changes

Giant slalom: compared to slalom there are higher speeds of up to 80 km/h and larger radii. As a result, the turn exposure time increases, but not the sum of acting forces. The competitions last about 80 s. Specific active prevention criteria:

- Optimization of mental and motor agility using training methods from the fields of coordination training or Life Kinetik® [17]
- Training of springiness under consideration of high contraction speeds and great holding forces
- Optimization of the interaction between eccentric and concentric motion sequences
- Specific training of endurance and force perseverance

- Extremely high movement precision due to the specific geometry of giant slalom skis that are moved at their absolute performance limit during turns

Speed disciplines (Super-G, Downhill): extreme alertness is needed here due to the changing terrain and the different sight and light conditions in combination with high speed.

The high speeds of up to 120 km/h expose the athletes to great forces of up to 7000 N [3] on turny course segments and landing after jumps. Specific demands:

- Optimization of risk management regarding the fact that competition conditions can frequently not be simulated in training. Course lengths, course preparation, and technical requirements are different in every race. Still, the race courses cannot be used for training. Considering the aforementioned criteria, the training courses are distinctly less demanding than the race courses. The athletes must be acquainted with competition conditions using speed elements. An important prevention measure is the athlete's perception of a manageable competition situation.
- Training of springiness under consideration of high contraction speeds.
- Optimization of the interaction between eccentric and concentric motion sequences.
- Specific training of endurance and force perseverance.
- Torso-stabilizing training with regard to the compression forces in the spine with neutralized S form.

10.4 Specific Preventive Activities in Training and Competition in Various Disciplines of Alpine Skiing

As can be quoted from the above accident statistics, more than 50 % of all injuries are located on the lower limb. The knee, followed by the hip and ankle, is most commonly involved.

Active joint stabilization focusing on the myofascial structures has priority. Contemporary scientific findings show the fasciae in a key role. The medical-therapeutic approaches have widened during the recent years through increasing scientific research [38].

Our whole body is composed of different types of connective tissue or fasciae. They surround, sheath, connect, separate, protect, and subdivide our body. The fasciae form a contiguous network of force transmission over the whole body [42]. Recent imaging shows microscopic as well as macroscopic supraregional connections.

The anatomic consideration of a global fascia network is an alternative to the anatomic insertion-origin classification with force distribution over one to maximally two joints. Classification of working groups or fascia chains has meanwhile been published by several authors [22, 23].

Proper muscle function is based upon its support and sheath function as well as the gliding capacity of the surrounding tissue layers. There is evidence for routes of different fascia chains that are anatomically connected with each other. Effective force transmission is the myofascial chains' main function. Thereby they form an inseparable entity with the muscles.

They can be regarded as cords intended to convey forces onto the body longitudinally or transversely, the joints serving as deflection pulleys [27] (Fig. 10.8).

There are four superordinate fascial layers: the superficial fascia covers the torso as well as the

Fig. 10.8 Myofascial chains are intended for effective force transmission. They form an inseparable entity with the muscles. The joints serve as pulleys

extremities and consists of fatty and loose connective tissues.

The profound fascia forms aponeuroses, peritendineum, periosteum, muscle sheaths, retinacula, etc. For example, the fascia lata extends into the crural fascia, again reaching into the plantar fascia.

The meningeal fascia coats the nervous system and, thus, forms the sheaths for the spinal cord.

The visceral fascia lines the body cavities and covers the internal organs. Consideration of these aspects yields a sophisticated perception of injury prophylaxis and should be integrated into prevention programs [28].

Dysfunctions of the fascial system: systemic processes as well as regional changes can appear in different body regions. Trigger points as fascial dysfunction are an example [12].

Fascial muscle dysfunctions result from various influences, such as immobilization, inflammation, pathological mechanical load, chronic repetitive overuse, hormonal derangement, and metabolic dysfunction or injury [1], in some cases with fascial distorsions [35]. In this context pathological alterations can develop in fascial structures. Classification of working groups or fascia chains has meanwhile been published by several authors [22, 34].

10.4.1 Preparative Measures

Severe changes should be diagnosed by a physician or therapist and treated as necessary. Among these, for example, are recurrent dysfunctions such as myotendinous irritations, acute low back pain, rheumatic conditions, various systemic diseases, and musculoskeletal disorders. A directed mechanical stimulus to the fascia chains to be trained as personal contribution has proven useful. The type of stimulus can be individually adapted. Pressure and traction, connective tissue rolls (Fig. 10.9), plucking, cupping glasses (Fig. 10.10), moving the skin in different directions, extensive self-massage, local compression, or cross-fiber friction (Figs. 10.11 and 10.12) should be applied here. Exercises with the black roll have shown to be very efficient. In doing so, the focus should be on slow movement (Figs. 10.13, 10.14, 10.15 and 10.16).

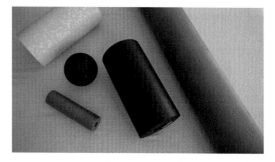

Fig. 10.9 Different types of fascia rolls (connective tissue rolls)

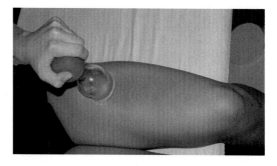

Fig. 10.10 Cupping glass massage of the quadriceps femoris muscle. Mechanical stimulus: traction

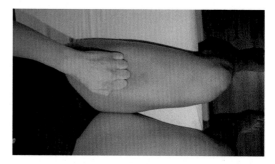

Fig. 10.11 Self-massage of the quadriceps femoris muscle. Mechanical stimulus: pressure

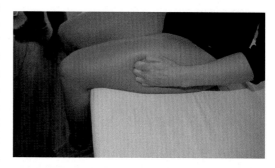

Fig. 10.12 Self-massage of iliotibial band. Mechanical stimulus: pressure

Fig. 10.13 Black roll massage of the quadriceps femoris muscle

Fig. 10.14 Black roll massage of the calf muscles

Fig. 10.16 Black roll massage of the paravertebral muscles

Fig. 10.15 Black roll massage of the iliotibial band

10.4.2 Characteristics of Myofascial Action and Collagen Fibers

The characteristics of myofascial action cannot be reduced to contraction and lengthening. Instead, the fasciae are able to absorb energy and release it, which can be attributed to their elastic properties. This should be considered in training. The collagen fiber is very adaptive. It changes according to the respective demands. Different qualities of collagen fiber enable adapting to repetitive loads (length, thickness, movability) [36]. Moreover, healthy fasciae show a so-called rebound effect:

elastically stored energy can be released rapidly, whereas this energy is commonly greater than the contractile force of the working musculature.

10.4.3 Fascial Training Under Consideration of Alpine Skiing Demands [30]

- Preparative counter motion: the activity to be performed is begun with an initial motion in the opposite direction.
- Movement execution should be smooth and soft. Changes of direction are carried out rhythmically.
- Dynamic stretching: flowing, swaying movement into the stretch.
- Reprogramming of neuromuscular proprioception: fascial receptors are trained for position and motion feedback.
- Hydration and regeneration: contraction and relaxation of the fasciae enhance the tissue fluid supply to the cells (mind fluid balance!!) [14].

Fig. 10.17 (**a**, **b**) Side-to-side swaying with crossed legs

Our recommendation for training duration and intensity:

Overall duration 20 min; 2–3 times per week; repeats, individual.
Exercise examples are given in Figs. 10.17, 10.18, 10.19, 10.20, 10.21, 10.22, 10.23, 10.24, 10.25, and 10.26.

10.5 Prevention of Overload and Reducing the Risk of Injury in High-Performance Skiing

The recreational athlete essentially has the possibility to avoid overload. Due to the performance density and present-day demands, this is not possible in competitive sports. To prevent overload damage in competitive sports, it is necessary to accelerate regenerative processes using thera-

Fig. 10.18 (**a**, **b**) Cat's arched back with rotation

Fig. 10.19 (**a–c**) Windmill forward-backward

peutic measures. This goal can only be achieved with intensive cooperation between trainers and therapists. In this regard mental prostration must also be considered. Qualified coaches realize the complex causes of fatigue. One substantial issue is whether the athlete is experiencing eustress or dys-stress. Hence, it is important to recognize the athlete's strain in training. An essential auxiliary tool in avoiding chronic overuse is putting variation into training, such as diverting leisure activities.

Hot spots concerning overuse and prevention in competitive alpine skiing are:

- The coach must know the athlete's personality.
- Great personal maturity of all involved is necessary.
- Pressure to perform must be analyzed and moderated.
- Athlete's preexisting physical conditions must be known.

- The coaches must be able to assess the athlete's presenting physical symptoms.
- In conclusion, overload must not lead to chronic overuse.

Concerning risk management in competitive sports, it is important in training to push limits, to

Fig. 10.20 (**a**, **b**) Diagonal flexion/extension with rotation while standing

recognize hazards, and to learn how to avoid harm when errors have occurred. This is the essential active strategy for injury prevention.

Hot spots concerning injury prevention in competitive alpine skiing are:

- Coaches must know the athlete's risk disposition.
- Posed challenges must be solved.
- The athlete must learn to handle critical situations.
- Avoiding of errors in critical situations teaches dealing with borderline situations.
- Necessary behavior in case of riding errors and falls is trained.
- Video analysis is useful for objectification of own experiences during intuitive learning [11].

10.6 Different Aspects for Male, Female, and Adolescent Athletes

Risks for dysregulation of vitamin and mineral uptake have been detected explicitly in female athletes [19]. In the case of vitamin D and calcium deficiency, an increased risk of stress fractures was observed in female athletes. Reduced

Fig. 10.21 (**a**, **b**) Mobilization of dorsal and ventral trunk musculature using the "lumberjack exercise"

Fig. 10.22 (**a–c**)
Strengthening exercises:
side-to-side or forward-
backward hops with both legs
3x30

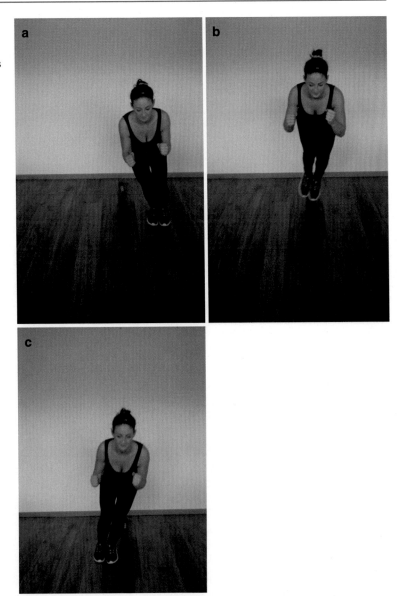

cognitive and physical performance is described in cases of iron deficiency. Greater blood loss resulting from menstruation is possibly the main cause for iron deficiency in premenopausal women and female athletes. Women with a large menstrual volume have a greater risk for poor iron status [13].

Noncontact knee injuries occur more often in women than in men and seem to be linked to the phases of the menstrual cycle. Presumably, female sex steroids can modulate ligament laxity via expression of relaxin, a peptide hormone. Elevated progesterone and estrogen levels stimu-late the relaxin receptor expression and thus increase connective tissue laxity [43]. The opposite effect is described for testosterone, which also promotes muscle growth. This is particularly seen in men receiving testosterone overdoses [7]. The majority of these results, though, originate from animal studies and can therefore only be transferred to humans to a limited extent [9]. There are also studies that show no correlation of ligament laxity and menstrual cycle [10].

In adolescent girls at an age of about 13 years, large knee abduction moments are the most probable cause for recurring retropatellar pain

Fig. 10.23 (**a**, **b**) Strengthening exercises: push-up 60°, dynamic with pushing off

Fig. 10.25 (**a**, **b**) Strengthening exercises: diagonal trunk exercises in supine position (abdominal muscles)

Fig. 10.24 Strengthening exercises: Downhill position static – teetering squats 3 × 90 s

Fig. 10.26 (**a**, **b**) Strengthening exercises: swimming exercise in prone position (back muscles)

and a greater risk of anterior cruciate ligament tears [21].

Due to children's physiology, overload syndromes in tendons and apophyses during growth pose a notable risk for the adolescent skier [2]. Consideration of resilience during individual growth phases requires great differentiation ability and empathy on the part of the coach. Overload damage is even described in 9-year-olds. Most commonly this kind of overload damage is seen in athletic 16-year-olds [20]. During alpine skiing, the knee extensors have a dominant role with patella overload, patellar tendinitis, Sinding-Larsen-Johansson syndrome (inflammatory reaction of the patellar tendon origin at the tip of the patella, occasionally with osteonecrosis, "jumper's knee"), and the Osgood-Schlatter disease (aseptic osteonecrosis of the tibial tuberosity). Some study results [4, 16] associate intensive physical strain, as in skiing, with pathological changes to the growing spine.

Yet, there is only limited knowledge about the long-term consequences of spine overload in the juvenile athlete, concerning morphological changes as well as causes for later back pain [6]. Ogon et al. [26] reported about radiologic surveys on 120 adolescent competitive skiers. They observed an increased risk of lower back pain only in subjects with distinct pathological ventral vertebral cover or baseplate alterations. No increased incidence was observed in adolescents with Schmorl's nodes.

Training intensity must be adapted to the respective development.

The coaches must recognize signs of fatigue in the child, mind rest, and nutrition. Overload syndromes of the musculoskeletal system must be prevented.

The most important active injury prophylaxis for children and adolescents in alpine skiing is sufficient physical and mental preparation [25]. Passive prevention in the form of protectors is becoming increasingly important. Symptoms of tendon irritation and muscle, joint, and back overload should be given due attention. Good endurance capacity is an essential prerequisite for optimal injury prophylaxis and performance in children's recreational and, even more so, competitive alpine skiing [44].

10.7   Passive Prevention

Passive prevention in skiers' equipment: apart from competitive skiing, passive prevention is becoming increasingly important in recreational skiing. Skiing helmets are presently worn by nearly all skiers. In competitive ski racing, the helmets are often complemented with further face protection. Helmets are partially required for children. Back protectors (Fig. 10.27) are the standard equipment for competitive athletes and are gaining acceptance in recreational skiing. In different disciplines of competitive skiing, upper and lower extremity protectors are becoming more prevalent. Clothing should be adapted to the weather conditions. Appropriate skiing goggles are important for identification of terrain and hazardous spots. Proper binding release setup is

Fig. 10.27 Back protectors are standard equipment for competitive athletes

Fig. 10.28 Preventive knee ortheses are being developed and partially used. Passive prevention in slope security: public information on weather conditions, weather forecasts, snow conditions, and avalanche warnings aid the skier's risk assessment. In competition different safety classifications for course security are specified by the FIS and constantly improved. Nets and air fences are mainly used. Slopes are increasingly secured with nets in recreational skiing

necessary to prevent lower extremity distortion injuries. Currently, preventive knee ortheses (Fig. 10.28) are being developed and partially used. Air bag systems for recreational and competitive skiing are being developed. Avalanche transceivers, avalanche backpacks with air bag, avalanche probes, and shovels are useful for backcountry skiing.

References

1. Aaron R, Bolander ME (2005) Physical regulation of skeletal repair. Symposium of the American Academy of Orthopaedic Surgeons, Washington D.C. 613–614
2. Adirim TA, Cheng TL (2003) Overview of injuries in the young athlete. Sports Med 33(1):75–81, Review
3. Babiel S, Hartmann U, Spitzenpfeil P, Mester J (1997) Ground-reaction forces in alpine skiing, cross-country skiing and ski jumping. In: Science and skiing. E&FN Spon (UK), London, pp 200–207
4. Bahr R, Andersen SO, Løken S, Fossan B, Hansen T, Holme I (2004) Low back pain among endurance
athletes with and without specific back loading – a cross-sectional survey of cross-country skiers, rowers, orienteerers, and nonathletic controls. Spine (Phila Pa 1976) 29(4):449–454
5. Bahr R, Krosshaug T (2005) Understanding injury mechanisms: a key component of preventing injuries in sport. Br J Sports Med 39
6. Baranto A, Hellstrom M, Nyman R et al (2006) Back pain and degenerative abnormalities in the spine of young elite divers: a 5-year follow-up magnetic resonance imaging study. Knee Surg Sports Traumatol Arthrosc 14:907–914
7. Bhasin S, Storer TW, Berman N, Callegari C, Clevenger B, Phillips J, Bunnell TJ, Tricker R, Shirazi A, Casaburi R (1996) The effects of supraphysiologic doses of testosterone on muscle size and strength in normal men. N Engl J Med 335(1):1–7
8. Bere T, Flørenes TW, Krosshaug T, Koga H, Nordsletten L, Irving C, Muller E, Reid RC, Senner V, Bahr R (2011) Mechanisms of anterior cruciate ligament injury in World Cup alpine skiing: a systematic video analysis of 20 cases. Am J Sports Med 39(7):1421–1429
9. Dehghan F, Muniandy S, Yusof A, Salleh N (2014) Sex-steroid regulation of relaxin receptor isoforms (RXFP1 & RXFP2) expression in the patellar tendon and lateral collateral ligament of female WKY rats. Int J Med Sci 11(2):180–191
10. Eiling E, Bryant AL, Petersen W, Murphy A, Hohmann E (2007) Effects of menstrual-cycle hormone fluctuations on musculotendinous stiffness and knee joint laxity. Knee Surg Sports Traumatol Arthrosc 15(2):126–132
11. Ettlinger CF, Johnson RJ, Shealy JE (1995) A method to help reduce the risk of serious knee sprains incurred in alpine skiing. Am J Sports Med 23:531–537
12. Gautschi R (2010) Manuelle Triggerpunkttherapie: Myofasziale Schmerzen und Funktionsstörungen erkennen, verstehen und behandeln. Thieme, Stuttgart
13. Harvey LJ, Armah CN, Dainty JR, Foxall RJ, Lewis DJ, Langford NJ et al (2005) Impact of menstrual blood loss and diet on iron deficiency among women in the UK. Br J Nutr 94:557–564
14. Klinger W, Schleip R (2007) Fascial strain hardening correlates with matrix hydration changes. In: Findles TW, Schleip R (eds) Fascia research – basic science and implications to conventional and complementary health care. Elsevier GmbH, München, S. 51 552–554
15. Kröll J, Spörri J, Müller E (2014) Verletzungsprävention im alpinen Skirennlauf – Eine wissenschaftliche Begleitung zur Einführung neuer Skispezifikationen, Bewegung & Sport. Heft 1/2014, S. 15–19
16. Kujala UM, Kinnunen J, Helenius P, Orava S, Taavitsainen M, Karaharju E (1999) Prolonged low-back pain in young athletes: a prospective case series study of findings and prognosis. Eur Spine J 8(6):480–484
17. Lutz H, Neureuther F (2009) Mein Training mit Life Kinetik: Gehirn + Bewegung = mehr Leistung. Verlag Nymphenburger, Muenchen

18. Manor M, Meyer N, Thompson J (2009) Sport nutrition for health and performance. Human Kinetics, Champaign
19. McClung JP, Gaffney-Stomberg E, Lee JJ (2014) Female athletes: a population at risk of vitamin and mineral deficiencies affecting health and performance. J Trace Elem Med Biol 28(4):388–392
20. Micheli L, Fehlandt AF (1992) Overuse injuries to tendon and apophysis in children and adolescents. Clin Sports Med 11(4):713–726
21. Myer GD, Ford KR, Di Stasi SL, Barber Foss KD, Micheli LJ, Hewett TE (2014) High knee abduction moments are common risk factors for patellofemoral pain (PFP) and anterior cruciate ligament (ACL) injury in girls: Is PFP itself a predictor for subsequent ACL injury? Br J Sports Med. doi:10.1136/bjsports-2013-092536. [Epub ahead of print]
22. Myers T (2009) Anatomy trains, 2nd edn. Churchill Livingstone, Edinburgh 457–462
23. Müller DG, Schleip R (2014) Lehrbuch Faszien. Urban und Fischer Verlag, Muenchen, 1. Auflage, S. 352–357
24. Natri A, Beynnon BD, Ettlinger CF, Johnson RJ, Shealy JE (1999) Alpine ski bindings and injuries: current findings. Sports Med 28:35–48
25. Neumayr G, Hoertnagl H, Pfister R, Koller A, Eibl G, Raas E (2003) Physical and physiological factors associated with success in professional alpine skiing. Int J Sports Med 24(8):571–575
26. Ogon M, Riedl-Huter C, Sterzinger W, Krismer M, Spratt KF, Wimmer C (2001) Radiologic abnormalities and low back pain in elite skiers. Clin Orthop Relat Res 390:151–162, Review
27. Paoletti S (2001) Faszien Anatomie. Strukturen. Techniken! 1. Auflage. Urban & Fischer, München, S. 178.f
28. Schleip R, Findley TW, Chaitow L, Huijing PA (2014) Lehrbuch Faszien. Urban und Fischer Verlag, Muenchen 1. Auflage. Allgemeine Anatomie der Muskelfaszie, S4.ff
29. Schleip R, Findley TW, Chaitow L, Huijing PA (2014) Lehrbuch Faszien. Urban und Fischer Verlag, Muenchen 1. Auflage. Allgemeine Anatomie der Muskelfaszie, S9–13
30. Schleip, Müller (2011) Faszien Fitness. Faszienorientiertes Training für Sport, Gymnastik und Bewegungstherapie. Terra Rosa E-Magazine, Issue Nr.7, S. 1–11
31. Schöllhorn W (2005) Differenzielles Lehren und Lernen von Bewegung. In: Göhner U, Schiebl F (eds) Zur Vernetzung von Forschung und Lehre in Biomechanik, Sportmotorik und Trainingswissenschaft. Czwalina, Hamburg, pp 125–135
32. Spörri J, Kröll J, Amesberger G, Blake OM, Müller E (2012) Perceived key injury risk factors in World Cup alpine ski racing – an explorative qualitative study with expert stakeholders. Br J Sports Med.; 46(15):1059–64
33. Stiftung Sicherheit im Skisport; Unfälle und Verletzungen im alpinen Skisport http://www.ski-online.de/stiftung-sicherheit/projekte/detail/asu-unfallanalyse.html
34. Tittel K (2003) Beschreibende und funktionelle Anatomie des Menschen. 14.A. Elsevier/Urban und Fischer, München
35. Typaldos S (1999) Orthopathische Medizin, Verlag für Ganzheitliche Medizin Dr. Erich Wühr, Bad Kötzting
36. Van den Berg F (2011) Angewandte Physiologie Band 1, Das Bindewebe des Bewegungsapparates verstehen und beeinflussen. Thieme, Stuttgart
37. Van den Berg F (2007) Angewandte Physiologie Band 3, Therapie, Training und Tests. Thieme, Stuttgart 330–332
38. Van der Wal JC (2009) Architecture of connective tissue as parameter for proprioception – an often overlooked functional parameter as to proprioception in the locomotor apparatus. Int J Ther Massage Bodywork 2:9–23
39. Van Wingerden JP, Vleeming A, Snijders CJ, Stoeckart R (1993) A functional- anatomical approach to the spine pelvis mechanism: interaction between the biceps femoris muscle and the sacrotuberous ligament. Eur Spine J 2:140–144
40. Waibel K, Spitzenpfeil P, Huber A; Techniktraining und Methodik, Rahmentrainingsplan Ski Alpin Deutscher Skiverband; http://www.dsv-datenzentrale.de/rahmentrainingsplan/3-TechniktrainingundMethodik-.htm
41. Waibel K, Spitzenpfeil P, Huber A; Sportmotorisches Anforderungsprofil, Rahmentrainingsplan Ski Alpin Deutscher Skiverband, http://www.dsv-datenzentrale.de/rahmentrainingsplan/52-Sportmotorisches_Anforderungsprofil-,e_508,r_47.htm
42. Willard FH (2007) The muscular, ligamentous and neural structure of the sacrum and its relationship to low back pain. In: Vleeming A, Mooney V, Stoeckert R (eds) Movement, stability and lumbopelvic pain. Elsevier, Edinburgh, pp 7–45
43. Winn RJ, Baker MD, Merle CA, Sherwood OD (1994) Individual and combined effects of relaxin, estrogen, and progesterone in ovariectomized gilts. II. Effects on mammary development. Endocrinology 135(3):1250–1255
44. Zintl F (1988) Ausdauertraining, BLV Buchverlag,. München

Implementation of Prevention in Sports

11

Stefano Della Villa, Margherita Ricci,
Francesco Della Villa, and Mario Bizzini

Key Points
1. The association between injuries and performance is one of the most important messages to deliver to coaches in order to enhance prevention.
2. A variety of injury prevention exercise programmes (IPEPs) have been shown to be effective in reducing the injury rate.
3. Implementation of injury prevention programmes represents a real challenge in sports medicine.
4. The coach is the key person to promote injury prevention to his/her players.
5. Prevention should be applied to all the players (general prevention) and to players at risk to be injured (tailored prevention).

11.1 How Do We Make Prevention Attractive?

Over the past 10–15 years, injury prevention has received a lot of attention in sports medicine. Despite this, prevention is not attractive. Our duty is to change this paradigm and to convince coaches, athletes and sports administrators to use prevention programmes. In this paragraph, we will discuss some of the "tricks" to make prevention more attractive.

First of all, it is important to design prevention exercises as structured warm-up, including these programmes into training sessions and competition. This routine could be useful to ensure that the players use the programme regularly [31].

We firmly believe that it will be easier to motivate athletes and coaches if such prevention exercises are shown to improve performance. Preventing injuries and therefore reducing the number of injured players means that the coach will have more players available for his/her ideal team. This has a strong impact in terms of the results of the team. A recent article by Hägglund et al. showed the importance of prevention to increase a team's chances of success [14]. Authors argued that the association between injuries and performance is one of the most important messages to deliver to coaches in order to enhance prevention [14]. In addition, the completion of The 11+ as a warm-up routine has been shown to help young futsal players to improve technical

S. Della Villa, MD (✉) • M. Ricci, MD
F. Della Villa, MD
Isokinetic Medical Group, FIFA Medical Centre of Excellence, v. di Casteldebole 8/4,
Bologna 40132, Italy
e-mail: s.dellavilla@isokinetic.com

M. Bizzini, PhD, PT
F-MARC (FIFA Medical Assesment and Research Centre), Schulthess Clinic,
FIFA-Strasse 20, Zurich 8008, Switzerland
e-mail: Mario.Bizzini@F-MARC.com

© ESSKA 2016
H.O. Mayr, S. Zaffagnini (eds.), *Prevention of Injuries and Overuse in Sports: Directory for Physicians, Physiotherapists, Sport Scientists and Coaches*, DOI 10.1007/978-3-662-47706-9_11

Fig. 11.1 The figure of the ambassador should be a coach or a famous player to implement the message worldwide

performance [28]. A correlation between lower incidence rate and team success has also been demonstrated by Eirale et al. [2].

Apart from preventing injuries and enhancing performance, other characteristics that make these programmes attractive are the requirement of minimal or no additional equipment and the fact that they are easy to do and do not take too much time [36].

A key aspect of successful injury prevention programmes is to include a variety of exercises with a progression from easy to more difficult to give a challenge to athletes. This is very important in order to motivate them. Also including exercises in pairs and a ball to make the training more fun could be useful [31]. Another fundamental issue is to involve players' feedback on injury prevention programmes for future directions as they are the real end beneficiaries [10].

We think that to be really successful, injury prevention programmes should be presented with these kind of messages and that players and coaches, rather than physicians, researchers and physiotherapists, would be the "ambassadors" (Fig. 11.1). We need to find the strategy to reach the community level.

11.2 Implementation of Prevention Programmes

Physical activity is associated with specific health benefits, including prevention of chronic diseases [38]. Injuries are indeed a common consequence of physical activity, and their management in sports, such as football (soccer), is difficult in terms of costs and time loss for the athlete, for the team and for the society as well [22]. It is known and well accepted that severe injuries take time to recover or can lead to an early retirement and to long-term consequences. For example, a majority of athletes sustaining an ACL injury will develop early osteoarthritis [23]. A variety of injury prevention exercise programmes (IPEPs) have been shown to be effective in reducing the injury rate [5, 16, 19, 20, 24, 29, 37]. Typically, these programmes consist of exercises focusing on balance, core stability, muscle strength and stretching. Examples of recent neuromuscular training (NMT) include a 5-phase 15 min programme [25], a 15-minute neuromuscular warm-up programme [37], the Prevent Injury and Enhance Performance programme (PEP) [24], The 11 [34] and The 11+ [29]. Although the effectiveness of these programmes has been demonstrated, compliance may be a critical factor in decreasing injuries [8]. To really prevent injuries and to have a significant public health impact, IPEPs need to be accepted, adopted and complied with by athletes, coaches, sports bodies and clubs [8]. Many studies demonstrated that players with high adherence/compliance had significantly lower injury risk [13, 30, 32]. In a recent meta-analysis, for example, Sugimoto et al. found that the incidence of ACL injury was lower in studies with good adherence to NMT [35].

Implementation of injury prevention programmes represents a real challenge in sports medicine [9]. The key point for future sports injury prevention is to reduce the "research to practice" gap [15]. It is mandatory to translate evidence into practice to establish the real effectiveness of these programmes.

As an example, we will discuss what the Fédération Internationale de Football Association (FIFA) has done in the past years and what it keeps on doing.

11.2.1 Implementation of the FIFA 11+ Football Warm-Up

The FIFA Medical Assessment and Research Centre (F-MARC), founded in 1994, has worked extensively in the development, scientific evaluation and dissemination strategies of FIFA's injury prevention programmes (i.e. FIFA 11+) [1]. The FIFA 11+ is a simple, and easy-to-implement, sports injury prevention programme (Fig. 11.2). It is a 20 min complete warm-up programme consisting of 15 single exercises, divided into three parts including initial and final running exercises with a focus on cutting, jumping and landing techniques (parts 1 and 3) and strength, plyometric, agility and field balance components (part 2). For each of the six conditioning exercises in part 2, The 11+ offers three levels of variation and progression [29]. Authors demonstrated a rate injury reduction of 37 % during matches and 29 % during training sessions [29]. The programme should be performed at the start of each training session, for at least twice a week. A key point in the programme is to use the proper technique during all of the exercises.

Basically, all researches were conducted in amateur/recreational football, which accounts for about 99 % of the 300 million players worldwide [7]. The F-MARC team has gained experience during the years of dissemination of the injury prevention programmes, especially through the countrywide campaigns in Switzerland, New Zealand and recently Germany.

11.2.2 Development of a Dissemination Strategy

The coach is the key person to promote injury prevention to his/her players [33]. While the coach, especially at a low level (where no staff personnel is available), has to oversee various aspects in the training (e.g. fitness, tactics), it is important to raise his/her motivation to implement an injury prevention programme (just as one of the many components) with his/her team.

Information material on "FIFA 11+" was developed, produced and made available for coaches and players. The material includes a detailed manual, an instructional DVD, a poster, a website and a promotional booklet with DVD. The material is available in the four FIFA languages (English, Spanish, German and French) and can be accessed in www.F-MARC.com/11plus.

"FIFA 11+" is best taught to coaches in a workshop that includes theoretical background knowledge and moreover practical demonstration of the exercises.

For the countrywide campaign in Switzerland, "The 11" was integrated in the coaching education of the Swiss Football Association (Schweizerischer Fussballverband (SFV)) using a "teach the teacher" strategy. All instructor coaches of the SFV were educated by sports physical therapists on how to deliver the programme in their licensing or refresher courses. During a period of 4 years, 5000 licensed amateur coaches were subsequently instructed on how to perform "The 11" with their teams and received the information material. In 2008, 80 % of all SFV coaches knew the prevention campaign "The 11" and 57 % performed the programme. Teams performing "The 11" had an 11.5 % lower incidence of match injuries and 25.3 % lower incidence of training injuries than other teams [18].

The same strategy was used in New Zealand, where "The 11" was implemented as part of the "SoccerSmart programme".

11.2.3 Worldwide Dissemination of "FIFA11+"

In 2009, FIFA started the dissemination of the "FIFA 11+" in its 209 member associations (MAs). Based on the experience with the countrywide campaigns, a guideline on how to implement the "FIFA 11+" injury prevention programme at a larger scale in amateur football

Fig. 11.2 The "FIFA 11+" programme is a classic example of NMT programme

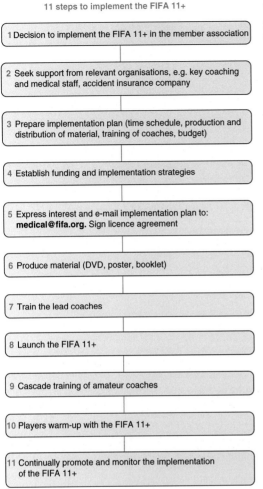

11 steps to implement the FIFA 11+

1 Decision to implement the FIFA 11+ in the member association

2 Seek support from relevant organisations, e.g. key coaching and medical staff, accident insurance company

3 Prepare implementation plan (time schedule, production and distribution of material, training of coaches, budget)

4 Establish funding and implementation strategies

5 Express interest and e-mail implementation plan to: **medical@fifa.org.** Sign licence agreement

6 Produce material (DVD, poster, booklet)

7 Train the lead coaches

8 Launch the FIFA 11+

9 Cascade training of amateur coaches

10 Players warm-up with the FIFA 11+

11 Continually promote and monitor the implementation of the FIFA 11+

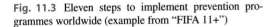

Fig. 11.3 Eleven steps to implement prevention programmes worldwide (example from "FIFA 11+")

was developed (Fig. 11.3). "FIFA 11+" has also been presented to the delegates of all MAs at the last two FIFA medical conferences (Zürich 2009, Budapest 2012). The implementation is conducted either in close cooperation with MAs or via FIFA coaching instructor courses. F-MARC supports the MAs in the preparation of the educational material in the local language and the workshops for the first group of instructors to initiate the cascade training. At MA level, it has to be acknowledged that highly motivated people are needed, in order to successfully plan, realise and constantly monitor a countrywide implementation.

The national Football Associations of Spain, Japan, Italy, Brazil and Germany integrated

"FIFA 11+" in their coaching curriculum and/or in their physical training/education curriculum. Thus, the world football champions took the lead and acted as role models, and other MAs (about 15 in November 2014) followed.

In the span of 5 years (2009–2014), FIFA 11+ has been presented in more than 80 countries worldwide (all continents), and thousands of coaches have been instructed on how to implement the programme, thus representing an important step for the worldwide dissemination of FIFA 11+ and the concept of injury prevention in football.

11.2.4 Socioeconomic Impact of Injury Prevention at Large Scale

The two countrywide campaigns in Switzerland and New Zealand represent successful examples of injury prevention in amateur football. Gianotti et al. introduced pre- and post-implementation cost outcome formulae to provide information regarding the success of a prevention programme [12]. These data provide a return on investment for each dollar invested in the programme and cost savings. Since the SoccerSmart programme (including the "The 11" programme) was introduced in New Zealand in 2004, the Accident Compensation Corporation (ACC) has invested 650,000 NZ dollars. Up to June 2011, ACC has saved 5,331,000 NZ dollars: the return of investment has risen to 8.20 for each invested dollar (personal communication of Dr. S. Gianotti, ACC, New Zealand). These data, together with the published results of the countrywide implementation in Switzerland, reinforce the hypothesis outlined by F-MARC back in 1994: prevention measures or programmes can not only reduce the incidence of football injuries but have the potential to save billions of dollars in health-related costs worldwide.

11.3   Target Groups for Prevention

Prevention programmes are thought to be applied in the same way to very different athlete groups. This approach is the one presented in the most

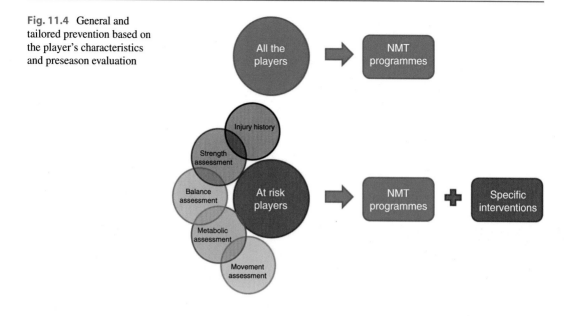

Fig. 11.4 General and tailored prevention based on the player's characteristics and preseason evaluation

part of articles. Whether or not a preseason evaluation based on movement analysis and athlete's clinical history will be useful has to be determined in further studies. Hewett et al. evaluated a large number of young female athletes in the preseason and built a predictive model for ACL injury risk based on a drop-jump analysis [17]. So the most logic approach may be to split in two the prevention concept: general prevention (through the well-known neuromuscular programmes) and tailored prevention (based on the player's risk profile) (Fig. 11.4).

In terms of general prevention, we strongly advice to implement prevention programmes on the two main categories of athletes: the recreational athletes and the young athletes.

If we really want to implement prevention and have a strong impact on public health, we need to target the recreational athletes (99 % of the 300 million football players worldwide). For recreational athletes we mean the competitive players of lower divisions (all sports) and the non-competitive recreational, maybe playing once a week without training. This is a hard challenge to be pursued. Regarding the first group, it is easiest to introduce prevention exercise as they play in a structured team with a dedicated staff or at least a coach, which has to be targeted for education. For the second group, public education programmes should be taught and implemented

together with physical activity prescription. Moreover, in case of injury, a specific reinjury prevention programme should be learned.

Most researchers and doctors recommend starting prevention strategies as soon as possible. Generally, injury risk is low under the age of 12 [11]. The incidence of injury increases with age, and studies have shown that the injury rate among players aged 16 years or older approaches that of adult players [3, 6]. The focus on young athletes ideally brings two main advantages. First of all, having a good prevention strategy at a young age means improving the young athlete's health also from a musculoskeletal perspective. Moreover, a proper education of the youngest means having a higher compliance of the future adult athletes. To reach the youngest, it is very important to involve the family and educate, first of all, their parents. Regarding the responsibility for implementing measures to reduce the risk of injury in the young population, Emery et al. argued that there is a hierarchy of responsibility with the lowest level of responsibility assigned to the child and the highest to those organisations (sports groups and government) that could potentially make the most change [4]. For example, an Australian legislation increased the helmet-wearing from 31 to 75 % in a year. This means that government regulation can make more changes compared to parent and child behaviour.

11.4 Ways of Publication of Prevention Programmes

First of all, the aim of an injury prevention research group is to check the effectiveness of the proposed NMT throughout prospective studies and subsequently share the findings with the scientific community. Among the 45 papers we reviewed during the writing of this chapter, we found that the most represented types of articles are prospective randomised controlled trial (RCT), prospective intervention study and statements. Prospective studies, possibly randomised, are the gold standard. An intervention and a control group need to be created. The control group may consist of a real group of different players without intervention or the same group of players in previous seasons.

Once the effectiveness of the NMT has been demonstrated, the scientific community has to deliver the programme to the public audience and to the sports community. This is what the international organisations, and in general all the sports bodies, are doing nowadays. Ways of "practical" publication have changed through the years, so, Web-based and app-based channels have to be carefully considered. Moreover, the organisation of workshops and courses for coaches and athletic trainers is still very important to spread out the message. The delivery of posters, DVD and pamphlets is fundamental as well. We need to use the right message to the right goal. For example, the right message to convince and motivate them is "injury prevention programme will reduce the risk by at least 50 %" instead of "injury prevention programme may reduce the risk" [26]. One of the keys while conducting a workshop or a course is "proposing" rather than "imposing" FIFA11+. After increasing the motivation of the coach and raising awareness of injury prevention, the exercises should be briefly explained and demonstrated. The participants should then perform the exercises and be corrected by the instructor(s), therefore appreciating the challenges (i.e. neuromuscular control!) behind each exercise. In the second half of the workshop, each of the participants should teach at least one of the exercises to the group and get feedback on this from the instructor. It is important to stress that regular and correct performance of the exercises is crucial for the preventive effect.

11.5 Importance of Media

Maybe the most important player on which we need to do prevention is the public audience. The sports medicine community needs the power of media and Web-based communication. Communication is evolving and the community is evolving too. Media can maximise the effort of a single man or organisation; the message may spread to the entire world and modify the way people think. This is what FIFA did during the World Cup in 2010 (Fig. 11.5), and they keep doing it through "The 11+" video sharing. Simple messages have to be transmitted to the audience.

Fig. 11.5 Prevention banner on the sideline as an example of implementation strategy

11.6 Duties of the Team Around the Player

Prevention has gained a lot of attention within the past years, and injury prevention is the responsibility of the whole team including the social athlete's environment, the sports federations, the coach, the medical staff and the athlete itself. To maximise the effectiveness of a sports-specific injury prevention programme, it is important to have this team approach.

11.6.1 Duties of the Social Environment of the Athlete

The social environment of the athlete consists of all the people involved in the life of the player. This group of people including the family, player manager, journalist and also supporters tends to give the athlete certain messages, directly or indirectly and with a different power in relation to the figure involved. These messages may modify the athlete's behaviour towards prevention. The social environment has a key role in guaranteeing the security and optimisation of the game in professional football, respecting the recent scientific evidence. Sustaining a major injury can lead to serious short-term and long-term consequences on the athlete's future health, including career ending, early retirement and chronic articular disorders, as knee or ankle osteoarthritis.

11.6.2 Duties of Sports Federations

Sports federations have the responsibility to minimise the risk of injury in all members. They should improve the level of knowledge of people taking care of players, through the organisation of courses for coaches and team staff.

The German experience has taught us that all these are possible through an allocation of resources and a strong central organisation. The German Football Association (DFB: Deutscher Fussballbund), the four-time World Cup winner, is the largest MA worldwide. The DFB has been for years a state-of-the-art organisation and has knowledge at any level of football: nevertheless, the Association decided in 2011 to promote "The

11+" NMT programme among its nearly 7 million registered amateur players. The 11+ was first presented to executives and representatives of the DFB amateur football, and in 2013 and 2014, more than 40 courses were conducted in the 21 regions of the DFB, for a total of more than 1000 coaches certified as FIFA 11+ instructors. These coaches will then be able to target the 26.000 registered clubs in DFB amateur football (a classical example of the "teach the teacher" strategy).

11.6.3 Duties of the Coach

The coach is the key person in the prevention programme and in the implementation process as well [26, 33]. Most of the time players trust their coach, so coach education is the real challenge of the prevention process. In a recent RCT, Steffen et al. argued that a preseason coach education workshop was more effective, in terms of better compliance and decreased injury risk in players, than an unsupervised Web-based delivery of The 11+. They also found that there is no additional benefit from the involvement of a physiotherapist [33]. We can conclude saying that the coach is the key partner ensuring high player adherence to the programme [26].

It has been found that understanding the coach's character and highlighting the importance of the programme to the coach are especially important. A primary goal of a coach at any level should be player development. Players who are injured are not training. Less training means less improvement and lack of opportunity to play at more competitive levels [21]. By preventing injuries and therefore reducing the number of injured players, the coach will have more players available for his/her ideal team. Additionally, studies have shown how the number of injuries correlates with the success and performance of the team [2, 14, 28].

The dialogue on the pitch with coaches ("speak the same language") is therefore often more important than the distributed materials, thus allowing for friendly discussion and practical work with the preventive programme. The choice of the instructors is crucial, and F-MARC's best experiences have been with sports physiotherapists or athletic trainers who have an active involvement in football. The cooperation with famous players and coaches acting as "FIFA 11+" ambassadors (see teaser on https://vimeo.com/45562029 and www.f-marc.com/11plus) has helped significantly in the communication with coaches.

11.6.4 Duties of the Medical Staff

Injury prevention (e.g. the ankle in football) is one of the major tasks of the medical staff including the physician and the physiotherapist. They have to stress the importance of prevention, especially towards young players. A good relationship between the coach and the players is a fundamental aspect, in which clear communication and mutual respect have a key role. They should also provide an individualised prevention programme based on the patient's risk profile. The interdisciplinary work with all the involved specialists in the team staff (e.g., fitness coaches and sports scientists) is important as well.

11.7 Literature Results

Most of the literature on prevention is about the efficacy of prevention programmes in football. From the implementation point of view, there are not too many studies, except from the ones by Donaldson and Finch.

In 2005, a pioneer RCT study by Olsen et al. was the first to demonstrate how a sports-specific injury prevention programme (as warm-up in the training routine) was highly effective in reducing the risk of noncontact knee and ankle injuries by half in youth female and male Norwegian handball [27].

The Soligard et al. study was the first large RCT showing how a structured injury prevention programme (FIFA 11+ developed within an international cooperation) could significantly reduce the risk of lower extremity noncontact injuries in youth female football players [29].

Junge et al. were the first to conduct and evaluate a countrywide campaign on injury prevention in football. Through coaching education, a simple programme (The 11) was disseminated and implemented in all amateur teams in Switzerland. The results showed an impact in reducing both injuries and healthcare-related costs [18].

In a recent RCT, Steffen et al. found how a preseason coaching educational workshop was the most efficient way to deliver an injury prevention programme (FIFA 11+), in terms of better compliance and even reduced injury risk (over one season) in youth female football teams [33].

Myklebust et al. describe some important lessons learned in 10 years of injury prevention in sports, underlying the importance of player's compliance and coach dedication in implementing an ACL injury prevention programme in Norwegian female handball [26].

References

1. Bizzini M, Junge A, Dvorak J (2013) Implementation of the FIFA 11+ football warm up program: how to approach and convince the Football associations to invest in prevention. Br J Sports Med 47(12):803–806
2. Eirale C, Tol JL, Farooq A, Smiley F, Chalabi H (2013) Low injury rate strongly correlates with team success in Qatari professional football. Br J Sports Med 47(12):807–808
3. Ekstrand J, Hägglund M, Waldén M (2011) Injury incidence and injury patterns in professional football: the UEFA injury study. Br J Sports Med 45(7):553–558
4. Emery CA, Hagel B, Morrongiello BA (2006) Injury prevention in child and adolescent sport: whose responsibility is it? Clin J Sport Med 16(6):514–521
5. Emery CA, Meeuwisse WH (2010) The effectiveness of a neuromuscular prevention strategy to reduce injuries in youth soccer: a cluster-randomised controlled trial. Br J Sports Med 44(8):555–562
6. Faude O, Junge A, Kindermann W, Dvorak J (2005) Injuries in female soccer players: a prospective study in the German national league. Am J Sports Med 33(11):1694–1700
7. FIFA. Big Count 2006 http://www.fifa.com/world-football/bigcount/ (cited 21 Feb 2013)
8. Finch CF (2006) A new framework for research leading to sports injury prevention. J Sci Med Sport 9(1–2):3–9
9. Finch CF (2011) No longer lost in translation: the art and science of sports injury prevention implementation research. Br J Sports Med 45(16):1253–1257
10. Finch CF, Doyle TL, Dempsey AR, Elliott BC, Twomey DM, White PE, Diamantopoulou K, Young W, Lloyd DG (2014) What do community football players think about different exercise-training programmes? Implications for the delivery of lower limb injury prevention programmes. Br J Sports Med 48(8):702–707
11. Froholdt A, Olsen OE, Bahr R (2009) Low risk of injuries among children playing organized soccer: a prospective cohort study. Am J Sports Med 37(6):1155–1160

12. Gianotti S, Hume PA (2007) A cost-outcome approach to pre and post-implementation of national sports injury prevention programmes. J Sci Med Sport 10(6):436–446

13. Hägglund M, Atroshi I, Wagner P, Waldén M (2013) Superior compliance with a neuromuscular training programme is associated with fewer ACL injuries and fewer acute knee injuries in female adolescent football players: secondary analysis of an RCT. Br J Sports Med 47(15):974–979

14. Hägglund M, Waldén M, Magnusson H, Kristenson K, Bengtsson H, Ekstrand J (2013) Injuries affect team performance negatively in professional football: an 11-year follow-up of the UEFA Champions League injury study. Br J Sports Med 47(12):738–742

15. Hanson D, Allegrante JP, Sleet DA, Finch CF (2014) Research alone is not sufficient to prevent sports injury. Br J Sports Med 48(8):682–684

16. Heidt RS, Sweeterman LM, Carlonas RL, Traub JA, Tekulve FX (2000) Avoidance of soccer injuries with pre-season conditioning. Am J Sports Med 28(5):659–662

17. Hewett TE, Myer GD, Ford KR, Heidt RS Jr, Colosimo AJ, McLean SG, van den Bogert AJ, Paterno MV, Succop P (2005) Biomechanical measures of neuromuscular control and valgus loading of the knee predict anterior cruciate ligament injury risk in female athletes: a prospective study. Am J Sports Med 33(4):492–501

18. Junge A, Lamprecht M, Stamm H, Hasler H, Bizzini M, Tschopp M, Reuter H, Wyss H, Chilvers C, Dvorak J (2011) Countrywide campaign to prevent soccer injuries in Swiss amateur players. Am J Sports Med 39(1):57–63

19. Junge A, Rösch D, Peterson L, Graf-Baumann T, Dvorak J (2002) Prevention of soccer injuries: a prospective intervention study in youth amateur players. Am J Sports Med 30(5):652–659

20. Kiani A, Hellquist E, Ahlqvist K, Gedeborg R, Michaëlsson K, Byberg L (2010) Prevention of soccer-related knee injuries in teenaged girls. Arch Intern Med 170(1):43–49

21. Kirkendall D (2013) Warm-up & injury prevention in football. ASPETAR Sports Med J 2:178–185

22. Lauersen JB, Bertelsen DM, Andersen LB (2014) The effectiveness of exercise interventions to prevent sports injuries: a systematic review and meta-analysis of randomised controlled trials. Br J Sports Med 48(11):871–877

23. Lohmander LS, Ostenberg A, Englund M, Roos H (2004) High prevalence of knee osteoarthritis, pain, and functional limitations in female soccer players twelve years after anterior cruciate ligament injury. Arthritis Rheum 50(10):3145–3152

24. Mandelbaum B, Silvers HJ, Watanabe DS, Knarr JF, Thomas SD, Griffin LY, Kirkendall DT, Garrett W Jr (2005) Effectiveness of a neuromuscular and proprioceptive training program in preventing anterior cruciate ligament injuries in female athletes: 2-year follow-up. Am J Sports Med 33(7):1003–1010

25. Myklebust G, Engebretsen L, Braekken IH, Skjølberg A, Olsen OE, Bahr R (2003) Prevention of anterior cruciate ligament injuries in female team handball players: a prospective intervention study over three seasons. Clin J Sport Med 13(2):71–78

26. Myklebust G, Skjølberg A, Bahr R (2013) ACL injury incidence in female handball 10 years after the Norwegian ACL prevention study: important lessons learned. Br J Sports Med 47(8):476–479

27. Olsen OE, Myklebust G, Engebretsen L, Holme I, Bahr R (2005) Exercises to prevent lower limb injuries in youth sports: cluster randomised controlled trial. BMJ 330(7489):449

28. Reis I, Rebelo A, Krustrup P, Brito J (2013) Performance enhancement effects of Fédération Internationale de Football Association's "The 11+" injury prevention training program in youth futsal players. Clin J Sport Med 23(4):318–320

29. Soligard T, Myklebust G, Steffen K, Holme I, Silvers H, Bizzini M, Junge A, Dvorak J, Bahr R, Andersen TE (2008) Comprehensive warm-up programme to prevent injuries in young female footballers: cluster randomised controlled trial. BMJ 337:a2469

30. Soligard T, Nilstad A, Steffen K, Myklebust G, Holme I, Dvorak J, Bahr R, Andersen TE (2010) Compliance with a comprehensive warm-up programme to prevent injuries in youth football. Br J Sports Med 44(11):787–793

31. Steffen K, Bahr R, Myklebust G (2010) ACL prevention in female football. ASPETAR Sports Med J 2:178–185

32. Steffen K, Emery CA, Romiti M, Kang J, Bizzini M, Dvorak J, Finch CF, Meeuwisse WH (2013) High adherence to a neuromuscular injury prevention programme (FIFA 11+) improves functional balance and reduces injury risk in Canadian youth female football players: a cluster randomised trial. Br J Sports Med 47(12):794–802

33. Steffen K, Meeuwisse WH, Romiti M, Kang J, McKay C, Bizzini M, Dvorak J, Finch C, Myklebust G, Emery CA (2013) Evaluation of how different implementation strategies of an injury prevention programme (FIFA 11+) impact team adherence and injury risk in Canadian female youth football players: a cluster-randomised trial. Br J Sports Med 47(8):480–487

34. Steffen K, Myklebust G, Olsen OE, Holme I, Bahr R (2008) Preventing injuries in female youth football – a cluster-randomized controlled trial. Scand J Med Sci Sports 18(5):605–614

35. Sugimoto D, Myer GD, Bush HM, Klugman MF, Medina McKeon JM, Hewett TE (2012) Compliance with neuromuscular training and anterior cruciate ligament injury risk reduction in female athletes: a meta-analysis. J Athl Train 47(6):714–723

36. Voskanian N (2013) ACL Injury prevention in female athletes: review of the literature and practical considerations in implementing an ACL prevention program. Curr Rev Musculoskelet Med 6(2):158–163

37. Walden M, Atroshi I, Magnusson H, Wagner P, Hägglund M (2012) Prevention of acute knee injuries in adolescent female football players: cluster randomised controlled trial. BMJ 344:e3042

38. Warburton DE, Nicol CW, Bredin SS (2006) Health benefits of physical activity: the evidence. CMAJ 174(6):801–809

Index

© ESSKA 2016
H.O. Mayr, S. Zaffagnini (eds.), *Prevention of Injuries and Overuse in Sports: Directory for Physicians, Physiotherapists, Sport Scientists and Coaches*, DOI 10.1007/978-3-662-47706-9